Springer Theses

Recognizing Outstanding Ph.D. Research

For further volumes:
http://www.springer.com/series/8790

Aims and Scope

The series "Springer Theses" brings together a selection of the very best Ph.D. theses from around the world and across the physical sciences. Nominated and endorsed by two recognized specialists, each published volume has been selected for its scientific excellence and the high impact of its contents for the pertinent field of research. For greater accessibility to non-specialists, the published versions include an extended introduction, as well as a foreword by the student's supervisor explaining the special relevance of the work for the field. As a whole, the series will provide a valuable resource both for newcomers to the research fields described, and for other scientists seeking detailed background information on special questions. Finally, it provides an accredited documentation of the valuable contributions made by today's younger generation of scientists.

Theses are accepted into the series by invited nomination only and must fulfill all of the following criteria

- They must be written in good English.
- The topic should fall within the confines of Chemistry, Physics, Earth Sciences, Engineering and related interdisciplinary fields such as Materials, Nanoscience, Chemical Engineering, Complex Systems and Biophysics.
- The work reported in the thesis must represent a significant scientific advance.
- If the thesis includes previously published material, permission to reproduce this must be gained from the respective copyright holder.
- They must have been examined and passed during the 12 months prior to nomination.
- Each thesis should include a foreword by the supervisor outlining the significance of its content.
- The theses should have a clearly defined structure including an introduction accessible to scientists not expert in that particular field.

Thomas M. Gøgsig

New Discoveries on the β-Hydride Elimination

Doctoral Thesis accepted by
Aarhus University, Denmark

Springer

Author
Dr. Thomas M. Gøgsig
Department of Chemistry
Aarhus University
Denmark

Supervisor
Prof. Dr. Troels Skrydstrup
Department of Chemistry
Aarhus University
Denmark

ISSN 2190-5053 ISSN 2190-5061 (electronic)
ISBN 978-3-642-32098-9 ISBN 978-3-642-32099-6 (eBook)
DOI 10.1007/978-3-642-32099-6
Springer Heidelberg New York Dordrecht London

Library of Congress Control Number: 2012943978

Printed on acid-free paper

Springer is part of Springer Science+Business Media (www.springer.com)

Parts of this thesis have been published in the following journal articles:

Studies on the Heck Reaction with Alkenyl Phosphates: Can the 1,2-Migration Be Controlled? Scope and Limitations. Ebran, J.-P.; Hansen, A. L.; Gøgsig, T. M.; Skrydstrup, T. *J. Am. Chem. Soc.* **2007**, *129*, 6931.

Studies on the 1,2-Migrations in Pd-Catalyzed Negishi Couplings with JosiPhos Ligands. Lindhardt, A. T.; Gøgsig, T. M.; Skrydstrup, T. *J. Org. Chem.* **2009**, *74*, 135.

Heteroaromatic Tosylates as Electrophiles in Regioselective Mizoroki-Heck Couplings with Electron Rich Olefins. Gøgsig, T. M.; Lindhardt, A. T.; Dekhane, M.; Grouleff, J.; Skrydstrup, T. *Chem. Eur. J.* **2009**, *15*, 5950.

Mild and Efficient Nickel-Catalyzed Heck Reactions with Electron-Rich Olefins. Gøgsig, T. M.; Kleimark, J.; Nilsson Lill, S.; Korsager, S.; Lindhardt, A. T.; Norrby, P.-O.; Skrydstrup, T. *J. Am. Chem. Soc.* **2012**, *134*, 443.

Carbonylative Heck Reactions Using CO Generated ex Situ in a Two-Chamber System. Hermange, P.; Gøgsig, T. M.; Lindhardt A. T.; Taaning R. H.; Skrydstrup T. *Org. Lett.* **2011** *13*, 2444.

Furthermore, seven scientific papers related to this work have been published and referenced throughout the thesis. Several of the papers have been highlighted by distinguished researchers. For example, the palladium 1,2-migration has been highlighted by Professor Ian J. S. Fairlamb (*Eur. J. Org. Chem.* **2009**, 4011) and Professor Zhang-Jie Shi (*Chem. Eur. J.* **2011**, *17*, 1728). The latter also covered the regioselective Heck reaction (Chapter 5), the iron catalyzed cross couplings (*Org. Lett.* **2009**, *11*, 4886), the nickel catalyzed Suzuki coupling of alkenyl phosphates (*Chem. Commun.* **2006**, 4137) and the Kumada coupling using ligand-free conditions (*J. Org. Chem.* **2009**, 74, 3536). In addition, the work regarding the direct vinylation strategy (*J. Org. Chem.* **2008**, *73*, 3404) has been highlighted by Professor Paul Knochel (*Synfacts* **2008**, 754) and Professor Scott E. Denmark (*Chem. Commun.* **2009**, 20).

Supervisors' Foreword

It is a great pleasure for us to write this introduction for truly an outstanding Ph.D. graduate, Dr. Thomas M. Gøgsig, who graduated from Aarhus University in November 2011. Thomas M. Gøgsig has been involved in the development of a number of significant transition metal-catalyzed reactions for the synthesis of organic compounds. Organic-based compounds are central for the economic growth of Europe, as chemical industries, which produce plastics, fuels, pharmaceuticals, rubber, cosmetics, detergents, coatings, dyestuffs, and agrochemicals, all rely on synthetic organic chemistry for their production. Most modern, high-tech materials are composed of organic compounds. Clearly, the availability of organic compounds produced by synthetic organic chemists is critically important for our high standard of living. An important area of organic chemistry is catalysis, which represents a very efficient means of synthesizing organic compounds. There are many different types of catalysts, but for those, which are composed of transition metals (38 elements in groups 3 through 12 of the periodic table), these represent some of the most important chemical entities for catalyzing the synthesis of organic molecules. The award of three Nobel Prizes in chemistry to this field in just the last decade truly reflects its importance (2001: For asymmetric metal-catalyzed reactions; 2005: For the development of metal-catalyzed metathesis reactions; 2010: For the discovery of palladium-catalyzed carbon–carbon bond forming reactions). Fundamental research from Academia is undoubtedly the main driving force for such discoveries as seven of the nine recipients of the Nobel Prize in catalysis are affiliated to a university. The unearthing of new elementary properties of transition metal based reagents has led to a greater comprehension of the reactivity of such catalysts, where its scientific outcome has in many cases provided new catalytic processes.

The scientific contributions from the Ph.D. work of Thomas M. Gøgsig falls specifically into this category. Thomas M. Gøgsig has worked in a very competitive area of expertise dealing with the discovery or improvement of a number of transition metal-catalyzed transformations, including (a) studies on Pd-catalyzed 1,2-migration reactions, (b) Pd-catalyzed Heck reactions employing heteroaromatic tosylates, (c) Ni-catalyzed Heck reactions, and (d) Pd-catalyzed carbonylation reactions employing ex-situ generated stoichiometric carbon monoxide. In all of this work, Thomas M. Gøgsig has shown excellent maturity as a chemist, dramatically improving the scope of the above-mentioned reactions and providing important mechanistic information that will, no doubt, help others to improve even further the outcome of these processes. Furthermore, the many publications from the work of Thomas M. Gøgsig have been published in leading international chemistry journals such as *J. Am. Chem. Soc.* and *Angew. Chem. Int. Ed.* In addition to these publications, two patent applications have also been submitted and Thomas M. Gøgsig has also been a key player in the setup of a new startup company SyTracks which is marketing some of Thomas M. Gøgsig's research discoveries. Certainly, the productivity of the thesis is highly impressive and many of his publications are key references within a very competitive field.

Prof. Troels Skrydstrup
Prof. John F. Hartwig

Preface

The author has been part of the OChem Graduate School Program, Aarhus University, and financially supported by the Faculty of Science. The work presented in this dissertation has primarily been performed at the Department of Chemistry, Aarhus University in the period August 2007–August 2011 under the supervision of Professor Troels Skrydstrup. The work presented in Chap. 4 relies on experiments conducted at the Chemical and Life Sciences Laboratory, University of Illinois at Urbana-Champaign, United States, under the supervision of Professor John F. Hartwig.

In Chap. 1, a presentation of the underlying principles of the organometallic chemistry involved in this dissertation is given.

Chapters 2 and 3 describe an extension of the palladium 1,2-migration in the Heck reaction discovered by Skrydstrup and coworkers. This work was performed in collaboration with Dr. Jean-Philippe Ebran and Dr. Anders Thyboe Lindhardt.

Chapter 4 describes the mechanistic studies on the palladium 1,2-migration in the Negishi and Kumada couplings, which were conducted under the guidance of Professor John F. Hartwig. The work has not been published so far.

In Chap. 5, a presentation of the related regioselective Heck reaction of heteroaromatic tosylates is given. The work was conducted in collaboration with Dr. Anders Thyboe Lindhardt, Dr. Mouloud Dekhane from AstraZeneca, and B.Sc. Julie Grouleff.

Chapter 6 describes a newly developed nickel-catalyzed Heck reaction. This work was performed in collaboration with Dr. Anders Thyboe Lindhardt and B.Sc. Signe Korsager. The computational part was conducted by the group of Professor Per-Ola Norrby; Dr. Jonatan Kleimark; and Dr. Sten Nilsson-Lill.

Chapter 7 describes the recently discovered carbonylative Heck reaction using a two-chamber system. The work was conducted in collaboration with Dr. Philippe Hermange, Dr. Anders Thyboe Lindhardt, and Dr. Rolf H. Taaning.

A representative selection of experimental details regarding the unpublished results presented in Chap. 4 and the nickel-catalyzed Heck reaction (Chap. 6) are also included in the appendices.

Denmark, July 2011 Thomas M. Gøgsig

Acknowledgments

First and foremost I would like to thank Professor Troels Skrydstrup for his guidance throughout the years and for the many hours of instructive discussions. I am very grateful to him for giving me the opportunity to work as a Ph.D. student and the liberty to pursue the questions generated by the research conducted in his laboratories. I would like to extend special thanks to my colleague and friend Dr. Anders Thyboe Lindhardt for the many successful collaborations and for proofreading this dissertation. In addition, I am very grateful to the former and present members of Group 107 for creating a relaxed but educational environment. I would like to highlight the following persons who have all participated and contributed to the research I have been involved in: Rolf H. Taaning, Jean-Philippe Ebran, Philippe Hermange, Delphine Gauthier, Lina Søbjerg, Signe Korsager, Dennis U. Nielsen, Klaus M. Bjerglund, Mette L. H. Mantel, Daniel Lupp, Kim L. Jensen, and Stephan Beckendorf.

I am deeply appreciative to Professor John F. Hartwig for allowing me to join his group for five months at the University of Illinois, Urbana-Champaign and for giving me the opportunity to learn from his expertise and comprehensive knowledge on organometallic chemistry. In this regard, I would like to thank the Hartwig Group for welcoming me and my wife. Thanks to Brad P. Carrow, Cass Richers, Christo Sevov, Seth Marquard, Alexey G. Sergeev, Ramesh Giri, Zheng Huang, David Huang, and especially Nan Holda.

Finally, I would like to express my sincere gratitude to my family at both ends of Denmark and for their kind help in times of need. Thanks to you Agnethe and for your untiring patience with me and to you Augusta for your sparkling smiles in the early mornings.

Acknowledgments

Contents

Abbreviations

Ac	Acetyl
9-BBN	9-Borabicyclo[3.3.1]nonane
BINAP	2,2′-bis(diphenylphosphino)-1,1′-binaphthyl
Bn	Benzyl
Boc	*tert*-Butyloxycarbonyl
bs	Broad singlet
calcd	Calculated
CBz	Carbobenzyloxy
COD	1,5-Cyclooctadiene
Cy	Cyclohexyl
Cy_2NMe	*N,N*-dicyclohexyl-*N*-methylamine
Cp	Cyclopentadienyl
d	Duplet
dba	*trans,trans*-dibenzylidene acetone
DBU	1,8-diazabicycloundec-7-ene
DFT	Density functional theory
DIPEA	Diisopropyl ethyl amine
DMF	*N,N*-dimethyl formamide
Dmphen	2,9-Dimethyl-1,10-phenanthroline
DMSO	Dimethylsulfoxide
D*i*PrPF	1,1′-bis(diisopropylphosphino)ferrocene
DPPPe	1,2-bis(diphenylphosphino)pentane
DPPE	1,2-bis(diphenylphosphino)ethane
DPPF	1,1′-bis(diphenylphosphino)ferrocene
DPPBz	1,2-Bis(diphenylphosphino)benzene
D*t*BPF	1,1′-bis(di*tert*-butylphosphino)ferrocene
EDG	Electron donating group
ee	Enantiomeric excess
EWG	Electron-withdrawing group
equiv	Equivalent
Fe	Iron

GC	Gas chromatography
h	Hour
HRMS	High resolution mass spectrometry
Hz	Hertz
i-Pr	Isopropyl
L	Ligand
LiCl	Lithium chloride
Mg	Magnesium
Me	Methyl
Mes	Mesityl
Met	Metal
min	Minutes
MOM	Methoxy methyl ether
n-Bu	*n*-Butyl
NHC	*N*-heterocyclic carbine
NMR	Nuclear magnetic resonance
Ni	Nickel
o	*Ortho*
o-tol	*Ortho*-tolyl
ORTEP	Oak Ridge Thermal Ellipsoid Plot
p	*Para*
Pd	Palladium
Ph	Phenyl
ppm	Parts per million
rt	Room temperature
s	Singlet
t	Triplet
t-Bu	*tert*-Butyl
TBAC	Tetrabutylammonium chloride
TBDMS	*tert*-butyldimethylsilyl
TBDPS	*tert*-butyldiphenylsilyl
TESOTf	Trimethylsilyl trifluoromethanesulfonate
Tf	Trifluoromethanesulfonyl
THF	Tetrahydrofuran
TOAB	Tetraoctylammonium bromide
Ts	*p*-Toluenesulfonyl
TS	Transition state
vs	Versus
X	Halide (or sulfonate)
Zn	Zinc

Ligands

cataCXium ligands: see p. 35

Q-Phos

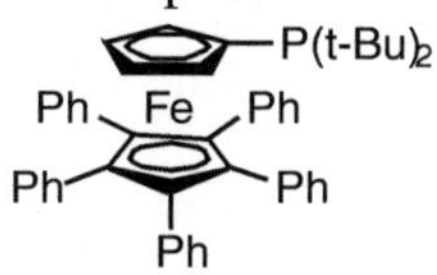

Josiphos family

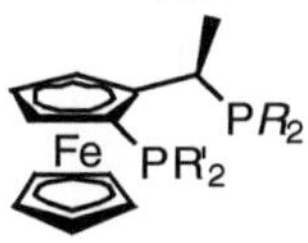

Buchwald ligands
(see p. 35 for examples)

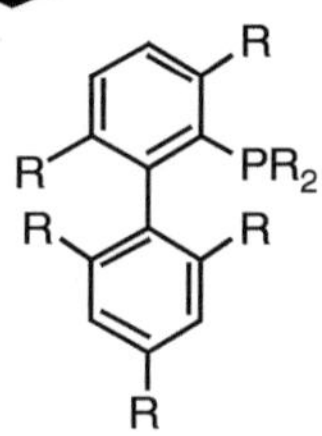

Chapter 1
Background

1.1 Introduction

Transition metal catalysis has since the pioneering work by Ullman in 1901 evolved into one of the most important means of facilitating organic chemical bond formations [1]. The introduction of transition metals opens for otherwise considered impossible transformations, and numerous strategies are thereby unlocked for the organic chemist. The applications are vast and spans from large scale productions e.g. the Monsanto acetic acid synthesis and the Wacker process, over highly advanced manipulations in the total synthesis of complicated natural compounds to academic research [2, 3]. The powers of transition metal catalysis are obviously immense, and therefore it is of outmost importance to clarify and understand the underlying mechanisms in order to optimize and further develop the huge potential of transition metals.

In particular, palladium has since its discovery in the early 1970s revolutionized the field of organometallic chemistry, and papers reporting new landmarks are published on a daily basis. During the last decades palladium catalysis has established its position as a very reliable method for the construction of C–C bonds. In this regard, cross coupling reactions have proved themselves extremely effective including the named reactions illustrated in Fig. 1.1 [4].

In the following sections a more detailed description of the individual elementary reactions, constituting the general catalytic cycles, will be presented. Focus will especially be directed towards the crucial β-hydride elimination exhibited by the Mizoroki–Heck reaction (Heck reaction) as this represents a central theme throughout the entire dissertation. In addition, examples on different reactions highly dependent on the β-hydride elimination will be given including some discoveries made in the Skrydstrup group, as these illustrate some elegant exploitations of an often undesired pathway.

T. M. Gøgsig, *New Discoveries on the β-Hydride Elimination*, Springer Theses,
DOI: 10.1007/978-3-642-32099-6_1,

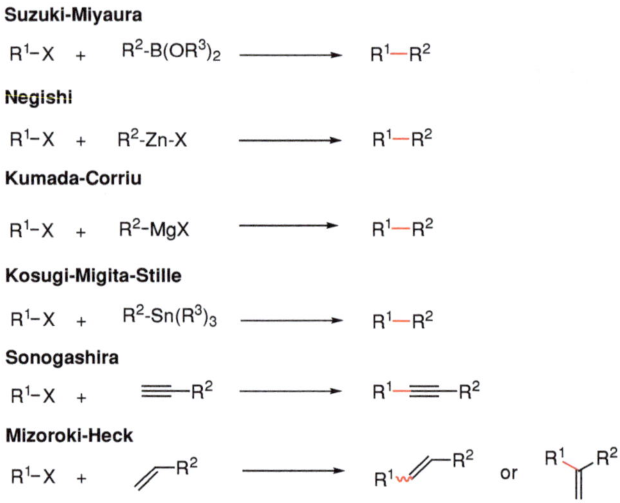

Fig. 1.1 Named Pd-catalyzed cross coupling reactions

1.2 The Active Palladium Catalyst

The efficiency of the catalytic reaction is typically highly dependent on the employed palladium precursor and the ligand. In most cases the precursor needs to be transformed into the active palladium(0) species prior to the initial oxidative addition and therefore lies off the catalytic cycle. This can be accomplished in a number of ways including ligand exchange, phosphine or base induced reduction, reductive elimination pathways etc. generating an unsaturated low valent palladium(0) complex (Scheme 1.1) [5–7].

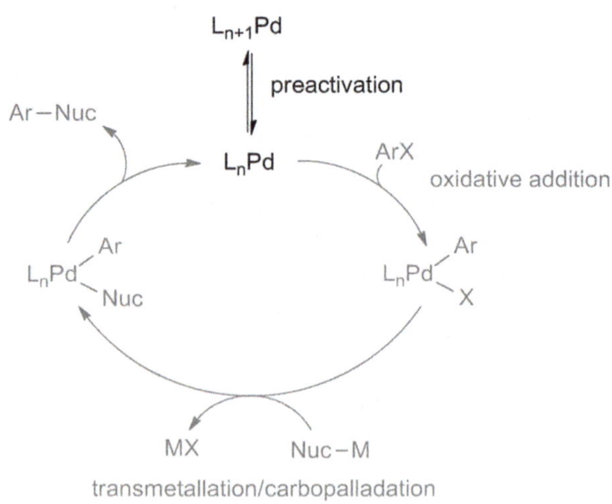

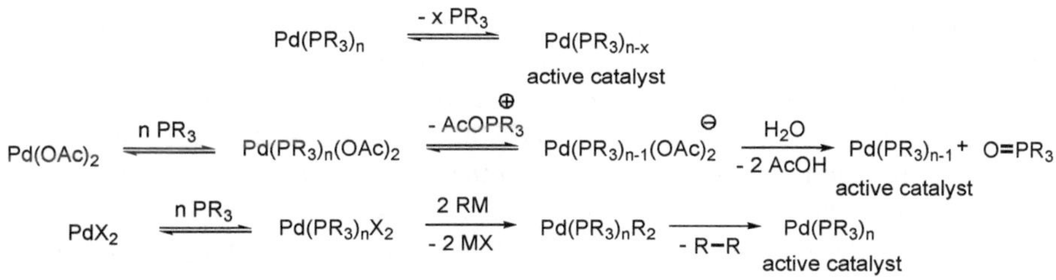

Scheme 1.1 Modes of catalyst activation

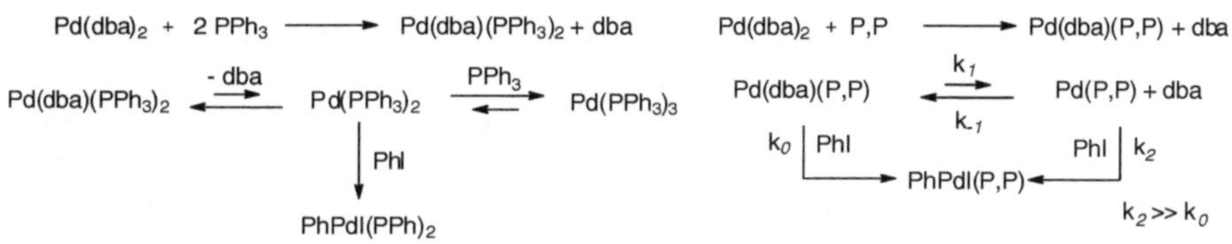

Scheme 1.2 Active catalytic species derived from $Pd(dba)_2$

The actual mode of activation relies on the particular catalytic process and can hugely influence the productivity of the overall reaction. Thus, a high concentration of the operative metal species, and hence a high yielding activation step of the catalyst precatalyst is desired. Traditionally, air and moisture stable palladium(II) salts such as $PdCl_2$ and $Pd(OAc)_2$ have been utilized as precatalysts. Alternatively, the palladium(0) complexes $Pd(PPh_3)_4$ and $Pd(dba)_2$ have been applied to a great extent representing a more direct approach to the active catalyst, although storage under an inert atmosphere is required for the former. The ligand displacement needed in order to generate the active catalyst from such complexes has been described and in some cases found ineffective, thereby potentially hampering the catalytic activity (Scheme 1.2) [8–10]. For example, the possibility of back donation into the π^*-orbitals of the chelating dibenzylideneacetone when using electron rich ligands has been observed, impeding the formation of the reactive palladium(0) species [11]. For further detail see Sect. 1.3.

Consequently, much attention has been directed towards the development of new and more readily activated precursors. Highly reactive bisphosphinepalladium(0) complexes such as $Pd[P(t\text{-}Bu)_3]_2$ and $Pd[P(o\text{-}tol)_3]_2$ more easily converted to the reactive catalytic species have proved to be efficient even in near homeopathic concentrations when combined with an appropriate ligand [12, 13]. The high levels of reactivity of these complexes towards oxidation demand the use of glovebox techniques, as these complexes decompose when subjected to atmospheric oxygen, which limits the attractiveness of such precursors. On the contrary, preligated catalysts based on palladacycles, [14, 15] palladium(I)-[16] or (II) [17] dimers or NHC-palladium(II) [18] complexes display enhanced stability and can often be stored on the bench for prolonged periods due to the higher oxidation state and saturated coordination sphere of the central metal. Such complexes are increasingly common and are often commercially available.

As mentioned, the ligands used in transition metal catalysis are of utmost importance and often the key to successful reaction. In principal, any group capable of interacting with the *d*-orbitals of the central metal can act as ligand. The ligands can be divided into two groups distinguished by their binding modes: covalent and dative ligands. The covalent ligands are covalently bound to the metal, thereby changing the oxidation state of the metal and are usually of negative charge. Examples include halides, carbanions, hydrides, alkoxides and so forth. On the other hand, the dative ligands do not change the oxidation state of the metal but deliver two electrons by π- or σ-bond donation to the metal *d*-orbitals and are of neutral character. The dative ligands are typically the ones considered when designing a catalytic system and are usually based on phosphines, carbenes, nitrogen containing compounds, π-systems etc [2]. In the following paragraphs phosphine ligands, representing the most widely used coordinating agents in palladium catalysis, will mainly be discussed in further detail. Nevertheless, many of the presented principles are valid for other dative ligands as well.

Phosphine ligands have since the discovery of palladium catalysis played a vital role in the development of effective catalysts. Today, a wide variety of both mono- and bidentate phosphine ligands are available displaying multiple sterical and electronic characteristics.

The steric properties of the two types of ligands are measured by the cone angle and bite angle, respectively (Scheme 1.3) [19, 20]. From Scheme 1.3 it can be envisioned, that a pronounced steric hindrance around the central metal is obtained by increased cone angle. As a consequence the active catalyst is in most cases monoligated during the catalytic cycle as reported by Hartwig and coworkers, when employing sterically encumbered monodentate ligands such as $P(t\text{-}Bu)_3$ [21]. In parallel, increased steric hindrance is obtained on the central metal, when bidentate ligands holding wide bite angles are introduced. Due to the intrinsically stronger chelating abilities displayed by the bidentate ligands, higher stabilities towards deactivation of the catalytically involved palladium complexes are often observed [22, 23]. On the other hand, metal complexes holding bisphosphine ligands tend to be less flexible than the corresponding monodentate ligated analogous. For example, complexes holding the covalent ligands *trans* to each other (*trans* isomers) cannot be formed unless bidentate ligands with large bite angles such as xantphos approaching 180° are employed [24]. The cone and bite angles of phosphine ligands can induce a great impact on the catalytic process and will be discussed in the sections regarding the elementary reactions constituting the catalytic cycle.

The electronic effects of the ligands can be adjusted by introducing different substituents on the phosphines, e.g. electron donating substituents on the phosphine result in better donating abilities of the phosphine. In particular, monodentate ligands containing trialkyl substituents have proved to be very potent providing a high electron density on the central metal. This in turn leaves the palladium very reactive towards the oxidative addition and C–X bonds

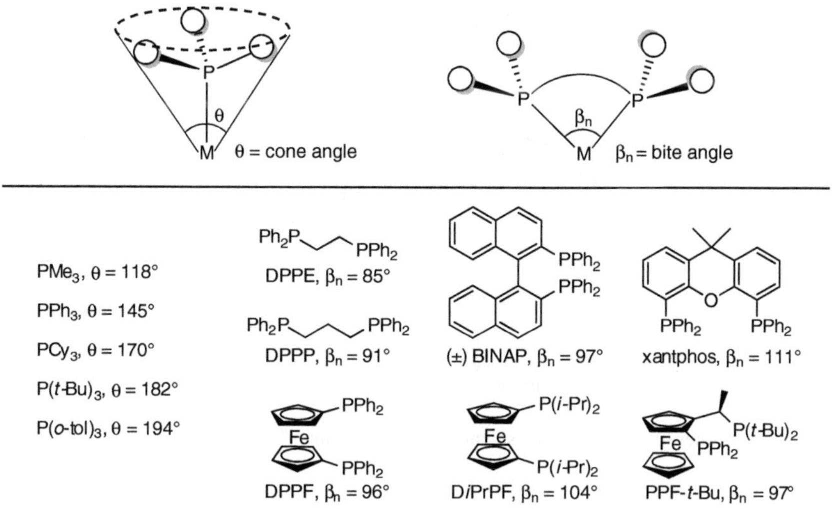

Scheme 1.3 Definition and examples of cone angle and bite angle

traditionally considered unreactive have been accessed with great success, including aryl chlorides representing a most valuable benchmark [25]. Unfortunately, such ligands are prone to oxidation and storage under an inert atmosphere is mandatory. This can however be circumvented by transforming the trialkyl phoshine into the corresponding HBF_4 salt as reported by Fu and coworkers [26]. The reactive phosphine is then formed in situ by adding a base.

Alternatively, the air- and moisture stable ligands introduced by the group of Buchwald holding a biaryl backbone have demonstrated exceptional donating abilities and have been utilized in variety of palladium catalyzed cross couplings including functionalization of aryl chlorides and mesylates [27, 28]. Furthermore, the Josiphos family combines the enhanced stabilizing effects of bidentate ligands and the electron donating abilities of alkyl substituents with a rigid ferrocene backbone. This results in highly stable catalytically active palladium complexes, securing excellent turn over numbers, even in the demanding activation of aryl tosylates [22, 29].

Even though the active catalyst has been produced effectively, a number of non-productive pathways during the catalytic cycle can be operative (Scheme 1.4) [2].

Reversible dimerization processes of intermediates involved in the productive steps and coordination of dative ligands rendering the metal unreactive are known to redirect the catalyst away from the catalytic cycle [30]. Potentially, pools of inactive catalytic species are created along the cyclic pathway lowering the concentration of the active catalyst and hence retarding the overall reaction. In addition, routes for irreversible deactivation such as C–P bond cleavage, cyclometallation of the coordinating ligand or metal precipitation may ultimately lead to catalyst depletion [31].

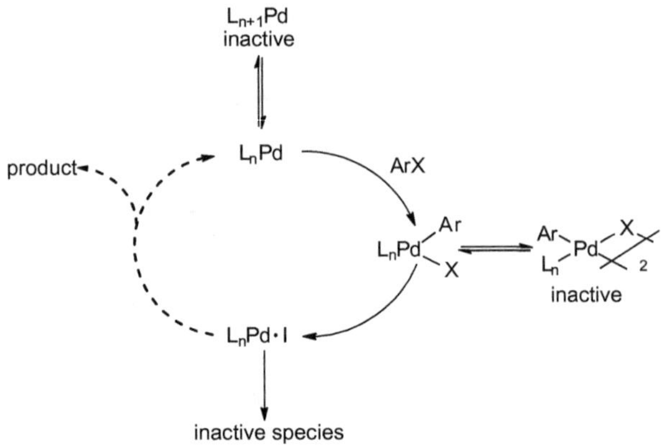

Scheme 1.4 Generation of inactive species during the catalytic cycle

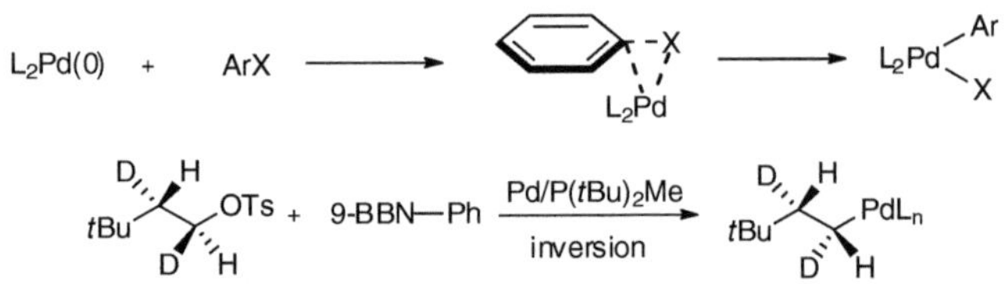

Scheme 1.5 Oxidative addition of C(sp^2)—X and C(sp^3)—X bonds to Pd(0)

The right combination of precatalyst and ligand is critical when designing a catalytic system. Intuition, based on thorough literature search, together with an extensive knowledge of the binding modes of the coordinating ligands can be decisive, although serendipity may be a significant factor during the initial screenings. The influence of the cone angle and bite angle of phosphine based ligands on the overall catalytic process is often hard to predict and may display opposing effects on the individual steps e.g. oxidative addition versus reductive elimination (see Sect. 1.3 and 1.6).

1.3 Oxidative Addition

The oxidative addition is the first reaction of the productive steps in the catalytic cycle and involves the addition of an electrophile to an unsaturated 14-electron bisphosphine or 12-electron monophosphine palladium(0) complex resulting in an oxidized palladium(II) species. The oxidative addition step is in principle reversible as the reverse reaction represents the reductive elimination, which will be discussed in Sect. 1.6 [2, 32]. The geometry of the central metal is altered from a tetrahedral palladium(0) compound to a square planar palladium(II) complex.

The mechanism underlying the oxidative addition of C(sp^2)—X bonds to palladium(0) complexes is considered to involve a three-centered concerted pathway, whereas the C(sp^3)—X bonds are activated in an S_N2 fashion by the reactive palladium(0) complex (Scheme 1.5) [33, 34].

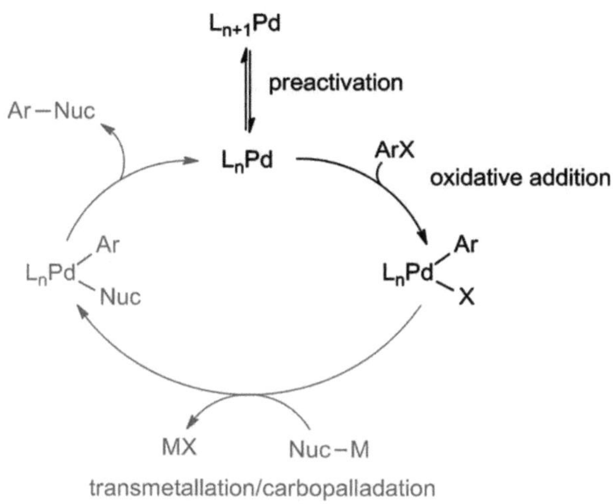

The electronic and steric effects of the phosphine ligands are crucial in the activation of aryl halides and pseudohalides. High electron density and steric bulk increase the capability of the palladium(0) species to undergo oxidative addition and account for the enhanced reactivity observed when exploiting monodentate ligands such as P(*t*-Bu)$_3$, cataCXium A, Q-phos and so forth. Due to the steric hindrance of these ligands only two phosphines are coordinated to the palladium(0) species. In many cases however, a dissociation of one of the phosphines is mandatory in order to generate the reactive palladium(0) species, which is favored by the increased steric effects displayed by this class of ligands. Even though a highly unsaturated monophosphine complex is generated, the electron donating properties of the ligand is still a vital factor (Scheme 1.6) [2, 35, 36].

As depicted in Scheme 1.6 the oxidative addition intermediate adopts a tricoordinated T-shaped geometry in contrast to the four coordinated square planar complexes formed, when less hindered phosphines are used (Scheme 1.7) [37].

The addition of anions to the reaction mixture may affect the rate of oxidative addition, creating a more polar medium or bind directly to the palladium and thereby changing the mechanism [38, 39]. The neutral bisphosphine palladium(0) complex is intrinsically more reactive than the anionic analogous due to the more "naked" central metal, but the concentration of the anionic species is significantly higher in polar media [2, 37]. Hence, the reaction will predominantly proceed through the tricoordinated anionic species, when polar solvents are used generating an intermediary pentacoordinated anionic palladium(II) complex.

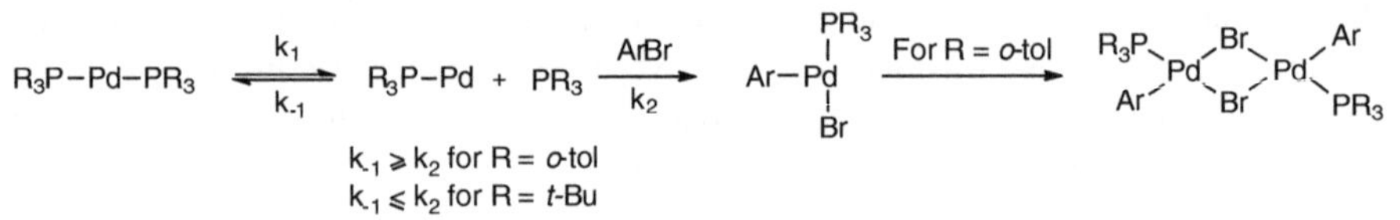

Scheme 1.6 Generation of active monophosphine Pd(0)-complexes

Scheme 1.7 Oxidative addition to bisphosphine Pd(0)-complexes

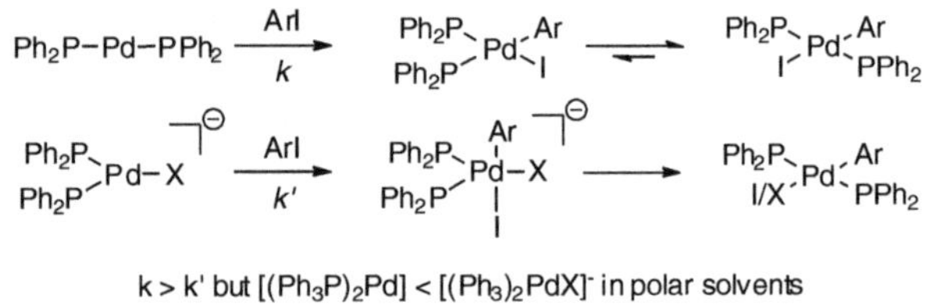

Scheme 1.8 Active complexes originating from linear and bend Pd(0)-complexes

The identity of the electrophile also affects the nature of the active catalyst when operating with extremely hindered monodentate phosphines. Accordingly, activation of aryl iodides, do not require an initial ligand dissociation, as they undergo oxidative addition to bisphosphine palladium(0) complexes, while the more reluctant aryl chlorides are only accessed through the more reactive monophosphine compounds [36]. Nevertheless, in both cases the tricoordinated T-shaped palladium(II) intermediate is formed.

The relative reactivities of the most common aryl halides and pseudohalides, regarding the oxidative addition, follow the trend ArI > ArOTf > ArBr > ArCl > ArOTs. However, these trends can be altered selecting an appropriate catalytic system. For instance, Fu and coworkers demonstrated that the combination of $Pd(dba)_2$ and $P(t\text{-}Bu)_3$ selectively reacted with ArCl in preference for ArOTf [40]. Potentially, this could be explained by a more stable oxidative addition complex obtained from ArCl compared to the ArOTf, comprising the weakly coordinating triflate counterion, provided that the oxidative addition is indeed reversible in such cases. Moreover, oxidative addition complexes obtained from aryl chlorides are known to dimerize when isolated, which could possibly stabilize the T-shaped triligated palladium(II) intermediate even further [36].

Oxidative addition to palladium(0) complexes containing bidentate ligands are obtained from unsaturated 14 valence electron bisphospine complexes. The bent conformation adopted by such complexes leaves the palladium(0) more reactive than the $L_2Pd(0)$ complexes obtained from monodentate phosphines with smaller cone angles, displaying a more linear geometry (Scheme 1.8) [2, 41].

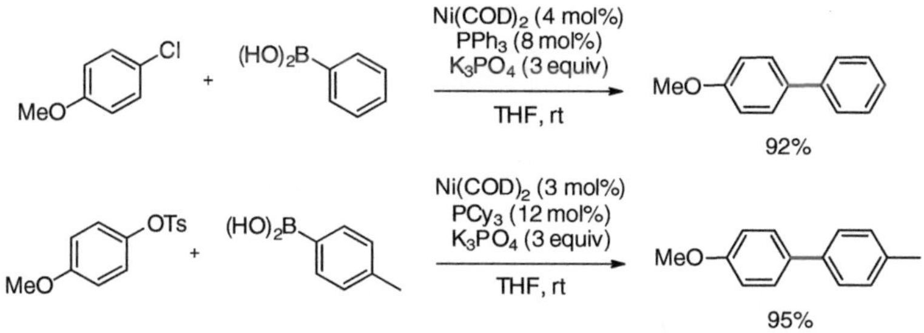

Scheme 1.9 Aryl chlorides and tosylates in Ni(0)-catalysis

The more available palladium(0) produced, when employing bidentate ligands holding a small bite angle, has been demonstrated to activate C—X bonds more readily than ligands with larger bite angles [42]. On the other hand, the ability of bidentate ligands to induce dissociation of other coordinating ligands, thereby creating the initially required 14-electron fragment, is significantly enhanced by increased bite angle of the bisphosphine [43]. Consequently, higher concentrations of the reactive palladium(0) species are obtained, when ligands with larger bite angles are used, forcing the other ligands away from the central metal to a greater extent, thus creating a more effective catalytic system. For example, Hartwig a coworkers reported aryl tosylates, usually considered inaccessible by bisphosphinepalladium(0) complexes, to undergo oxidative addition even at room temperature, when palladium complexes containing bidentate Josiphos ligands holding wide bite angles were evaluated [44]. However, reduced catalytic activity was in some cases observed, when $Pd(dba)_2$ was chosen as precatalyst, most likely due to an enhanced back donation from the initial (L—L)Pd(0) complex into the π^* orbitals of the dibenzylideneacetone hindering the crucial dissociation [11].

Alternatively, nickel(0) catalysts have proven to be more reactive towards the oxidative addition than the corresponding palladium(0) protocols, and aryl chlorides and tosylates are easily activated (Scheme 1.9) [45, 46].

Furthermore, more recent strategies relying on nickel based catalytic systems have described the functionalization of aryl ethers, carboxylates, carbamates and so forth, thereby dramatically increasing the scope of transition metal catalysis [47].

1.4 Transmetallation and Carbopalladation

The coordination and subsequent assimilation of the nucleophilic coupling partner to the discrete oxidative addition complex generating a new diorganopalladium(II) species represent the next step in the catalytic cycle. Depending on the type of the catalytic process and the incoming nucleophile two major pathways are typically considered; transmetallation or carbopalladation. In the following a short presentation and description of the two distinct reactions will be given.

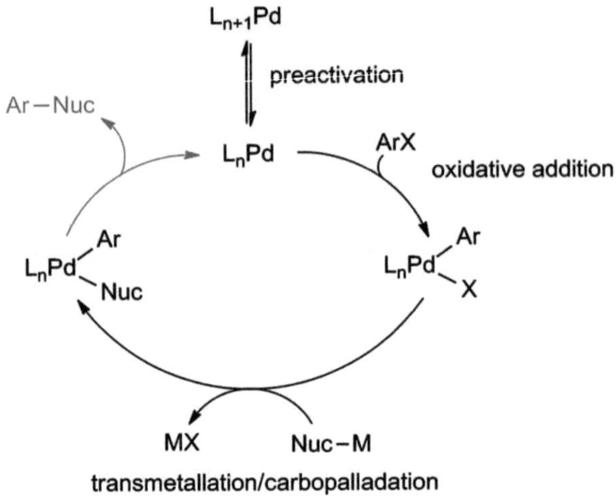

1.4.1 Transmetallation

Transmetallation refers in this context to the bimolecular transfer of an organic group from an organometallic species to a palladium(II) complex generating a new Pd—C bond and a M—X bond. The principle of transmetallation is exploited in the named cross coupling reactions depicted in Fig. 1.1 and traditionally involve organometallic species derived from the elements boron, zinc, magnesium, tin, silicon (Hiyama) and copper utilized in the copper promoted Sonogashira coupling. The exact mechanism of a general transmetallation is not clear, but it is anticipated that a four-centered transition state is operating (Scheme 1.10) [48].

The driving force of the transmetallation is believed to originate from the difference in electronegativity between palladium ($\chi = 2.20$) and the incoming organometallic species resulting in increased rates, when metals of low electronegativity such as magnesium ($\chi = 1.31$) are employed compared to for example boron derived nucleophiles ($\chi = 2.04$) [49]. However, organometallic species of low nucleophilicity tend to display an enhanced functional group tolerance as illustrated by the Suzuki, Stille and Negishi couplings [50]. In the Suzuki and Stille couplings, the addition of an activator such as base or a coordinating agent is often necessary. Additionally, studies on the organometallic species conducted by Stille and coworkers showed the following order of reactivity alkynyl > alkenyl > aryl > benzyl > methyl > alkyl towards the transmetallation in the Stille coupling of benzoyl chlorides [51].

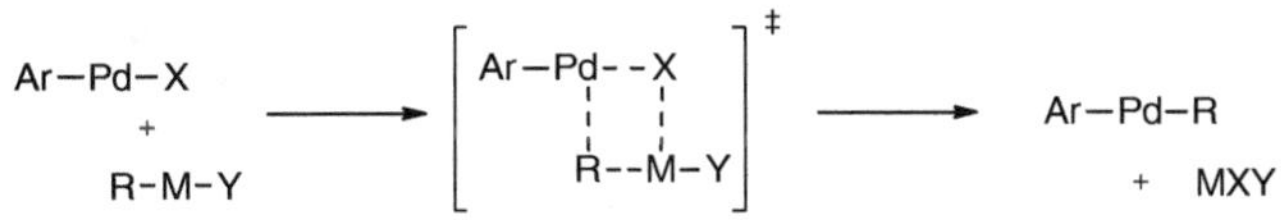

Scheme 1.10 Mechanism of transmetallation

Scheme 1.11 Coordination and carbopalladation

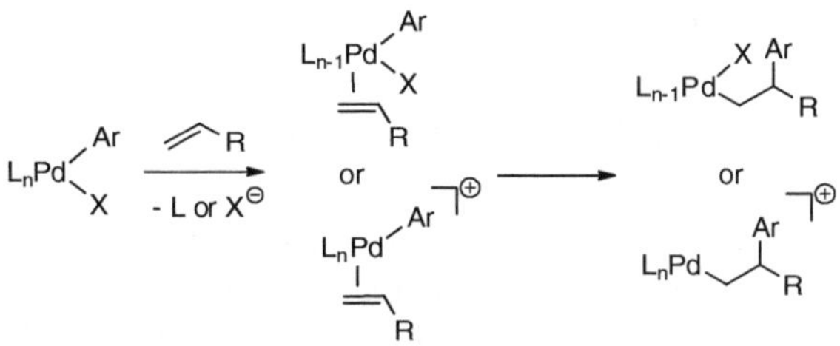

1.4.2 Carbopalladation

When olefins are employed as the coupling partner in the palladium catalyzed reactions, the overall catalytic cycle is altered compared to the cross coupling reactions of organometallic species. In parallel to some of the cross couplings a base is needed in order to generate an efficient catalytic cycle although the exact role is different. The palladium catalyzed olefination of aryl and vinyl halides and pseudohalides is referred to as the Heck reaction, due to the early discoveries made by Heck [52]. Later on, the beneficial role of added base, allowing the use of catalytic amounts of palladium, was independently reported by the groups of Mizoroki and Heck [53, 54]. In order to coordinate the olefinic moiety a ligand dissociation from the four coordinated square planar palladium(II) complex generating a vacant site is mandatory. Subsequently, *syn*-addition of the Pd–C to the double bond forms an alkylpalladium(II) intermediate as illustrated in Scheme 1.11.

Bidentate ligands are most often employed when a cationic palladium(II) intermediate is desired, whereas monodentate ligands are typically used in the generation of the neutral palladium(II) species [55]. On the contrary, a free coordination site is inherently available in the tricoordinated T-shaped palladium(II) intermediate obtained with highly electron donating and sterically encumbered ligands [56].

The successive coordination and *syn*-addition of organopalladium(II) intermediates to olefins and alkynes generating a new C—C σ-bond are referred to as migratory insertion or carbopalladation and control the regioselective outcome in the particular reaction. The regiochemistry can be tuned by steric and electronic effects, which will be discussed in more detail for the Heck reaction in Sect. 1.7.

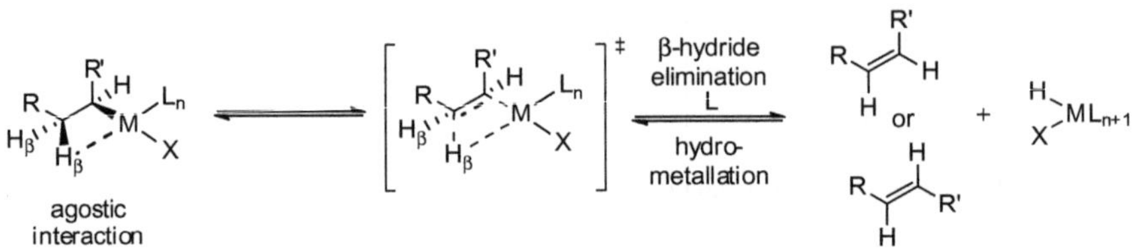

Scheme 1.12 Geometric requirements for *β*-hydride elimination

1.5 *β*-Hydride Elimination

The elimination of a hydrogen atom positioned on a carbon *β* to the central metal constitutes an important reaction in transition metal catalysis. In the classical example, an alkylmetal intermediate is reversibly converted to an alkene and a metalhydride (scheme 1.12). Despite the fact, that the resulting hydridometal complex can be exploited in various catalytic processes including polymerization reactions, [57] cycloisomerizations, [58] annulations, [59] etc., the *β*-hydride elimination is often considered undesired in transition metal catalyzed cross couplings. Thus, efforts have often been concentrated towards the prohibition of this fundamental reaction [60]. Nevertheless, the *β*-hydride elimination is a vital transformation in a number of catalytic processes including the ene-yne coupling reported by Trost [61] and Skrydstrup, [62] oxidation of alcohols, [63] the Heck reaction etc [64].

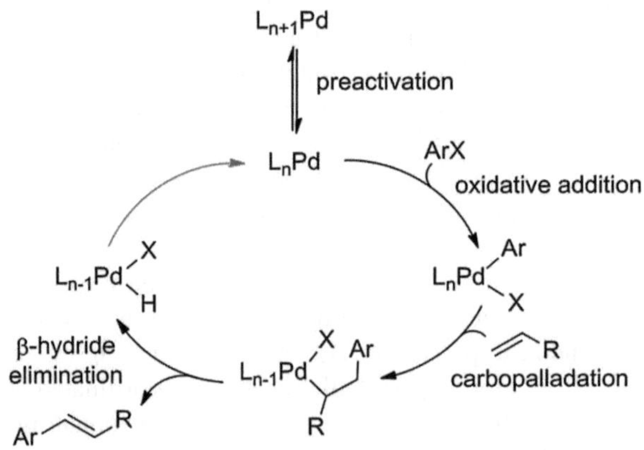

In order to facilitate the *β*-hydride elimination a few requirements have to be fulfilled. A free coordination site on the metal center allowing the migration of the hydrogen to the metal, is mandatory, i.e. a ligand dissociation from a saturated alkylmetal intermediate must occur initially. Furthermore, a *syn* coplanar arrangement of the hydrogen and the central metal is required. An agostic interaction between the hydrogen and the metal precedes the elimination reaction creating a four-centered transition state, which collapses into the resulting alkene

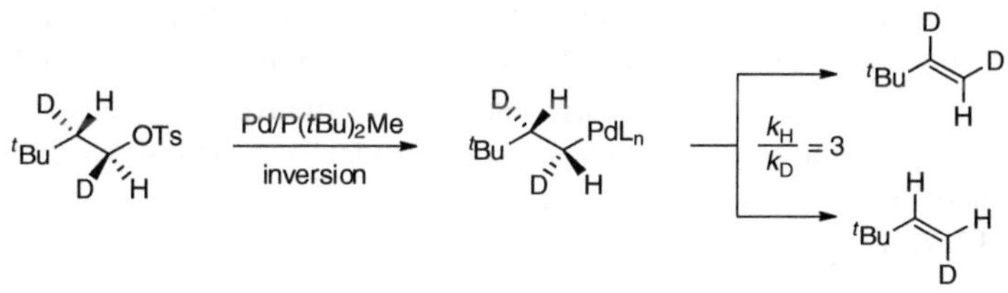

Scheme 1.13 Unimolecular β-hydride elimination

Scheme 1.14 β-hydride elimination from alkenyl intermediates

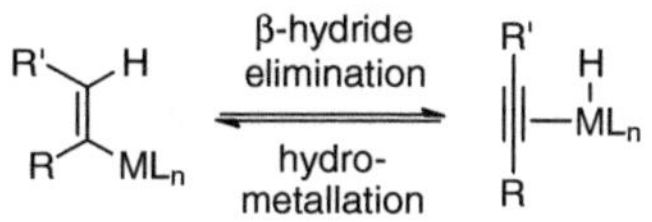

and a metal hydride complex. As illustrated in Scheme 1.12 the stereoselective outcome of the reaction is determined in this step. If the R-groups are different from hydrogen, the thermodynamically more stable alkylmetal complex, holding the R-groups *anti* to each other, is initially formed instead of the depicted *syn* complex, resulting ultimately in the *anti* alkene [65]. Evidence for a unimolecular mechanism has been provided by Fu and coworkers, demonstrating a small preference for the elimination of hydrogen compared to deuterium in the β-hydride elimination from an alkylpalladium(II) complex (Scheme 1.13) [34].

The β-hydride elimination from the corresponding alkenylmetal complex, forming an alkyne in addition to the metalhydride, has only been sporadically described in the literature indicating a much less feasible elimination process from such complexes. Hydrometallation, representing the reverse reaction, to alkynes is considered fast and could potentially account for the lack of data regarding alkenyl β-hydride eliminations (Scheme 1.14) [66].

The β-hydride elimination and its applications in the Heck reaction will be discussed in further detail in Sect. 1.7.

1.6 Reductive Elimination

The reductive elimination represents the reverse reaction of oxidative addition and constitutes the last elementary reaction in the general catalytic cycle. Catalyst regeneration and liberation of the desired cross coupling product are typically obtained at the same time during this step. Since the reductive elimination from transition metal complexes is the reverse reaction of the oxidative addition, involving the same transition states, the principles described in Sect. 1.3 prevail, although in opposite direction (Scheme 1.15). As a consequence, the underlying mechanism creating the new σ-bond between two covalent ligands bound to palladium is believed to involve the same concerted three-centered pathway regenerating the reduced metal species. A *cis* relationship between the two departing groups is required in order to accommodate the orbital overlap, which is only obtained by a stepwise mechanism from complexes possessing a *trans* configuration [2, 67].

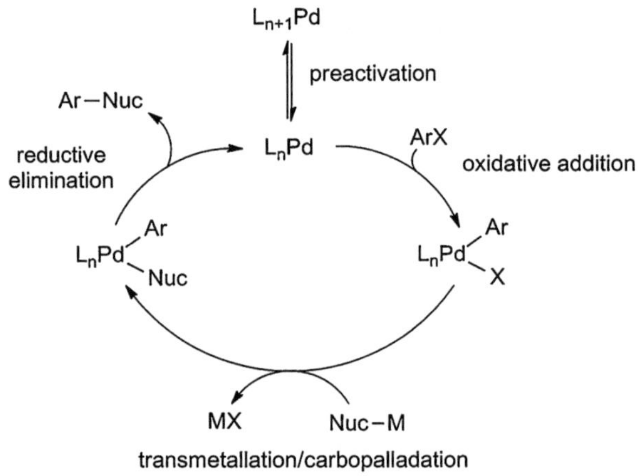

A number of factors are known to influence the reductive elimination from palladium(II) complexes. In contrast to the oxidative addition faster rates for the reductive elimination is observed from electron poor metal centers compared to the electron rich equivalents. Thus, highly electron donating ancillary ligands tend to retard the reductive elimination, stabilizing the palladium(II) intermediate. Nevertheless, when an odd numbered coordinated palladium(II) species is operational, the reductive elimination occurs more rapidly than the four- or sixcoordinated palladium(II) complexes [67, 68]. This effect can be explained by analysis of the frontier σ* orbitals of the metal—ligand bonds and accounts for the observed efficiency of the bulky and strongly electron donating trialkyl phosphines forming tricoordinated T-shaped intermediates [69].

Steric hindrance around the metal center elevates the overall energy of the palladium complex. During the reductive elimination and release of the final product, the coordination number is decreased by two, and hence the compiled energy of the sterically congested palladium(II) complex is released [67]. Consequently, applying bidentate ligands with large bite angles accelerates the reductive elimination, which has been demonstrated by Grushin reporting the demanding reductive elimination of trifluoromethylbenzene from a xantphospalladium(II) complex (Scheme 1.16) [70, 71].

On the other hand, DPPBz, containing a smaller bite angle, did not facilitate the formation of trifluoromethylbenzene even after days at elevated temperatures. In most cases, the steric effects have a larger impact on the rate of reductive elimination compared to the electronics properties of the coordinating ligands [69].

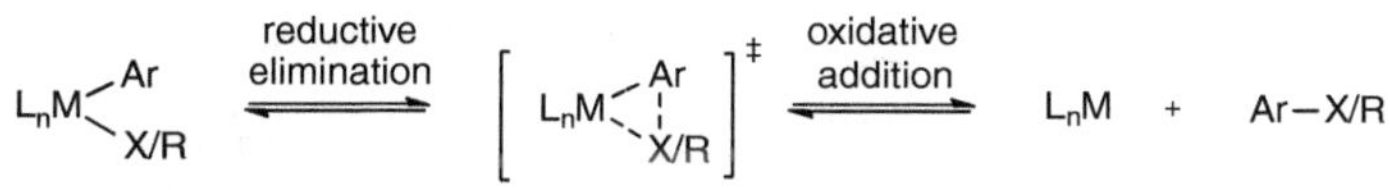

Scheme 1.15 Reversibility of reductive elimination and oxidative addition

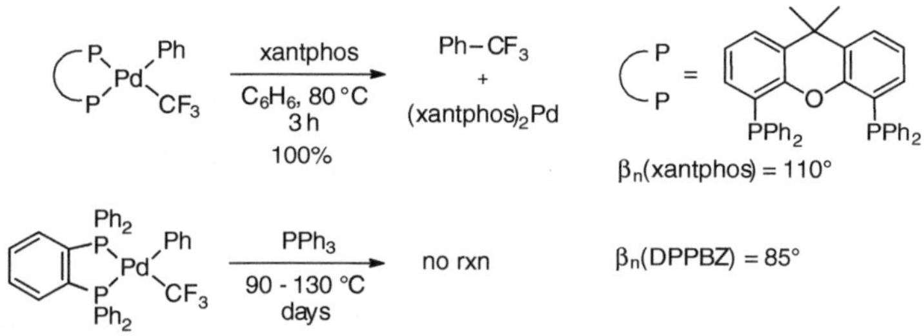

Scheme 1.16 Effects of steric properties on the reductive elimination

The electronic properties of the covalent ligands also influence the rate of the reductive elimination. Faster rates are observed, when covalent ligands carrying electron deficient groups are combined with the electron rich analogous compared to covalent ligands of similar electronic propensities. Furthermore, the hybridization of the carbon covalently attached to the palladium greatly influences the rate of catalyst regeneration obtaining increased rates, when aryl and vinyl groups are employed compared to the alkyl equivalents. These trends are approximate and can be altered by the applied reaction conditions but may act as initial guidelines [67].

Contrastingly, the Heck reaction the product is released by the β-hydride elimination and a palladium(II) hydride complex is formed. Regeneration of the catalyst is achieved by the assistance of a base. The base induced reductive elimination from hydridopalladium(II) complexes is less studied but most likely involves a coordination of the base, typically an amine, to the palladium hydride intermediate [72]. On the contrary, the base could potentially act as a proton scavenger trapping the HX released from the metal center by a unimolecular reductive elimination process and thereby regenerate the palladium(0) species. Studies conducted by Fu and coworkers on the stoichiometric reaction of $[P(t\text{-}Bu)_3]_2PdHCl$ in the presence of base suggested an initial dissociation of one of the ligands generating the tricoordinated T-shaped palladium hydride complex. This complex then reacts with the base and regenerates the palladium(0) complex. On the other hand, the less sterically hindered $(PCy_3)_2PdHCl$ did not undergo base promoted reductive elimination, indicating a significant sterical dependency (Scheme 1.17) [72].

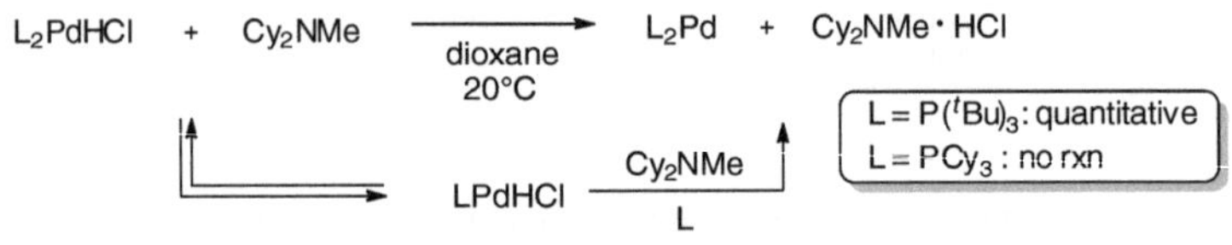

Scheme 1.17 Base induced reductive elimination

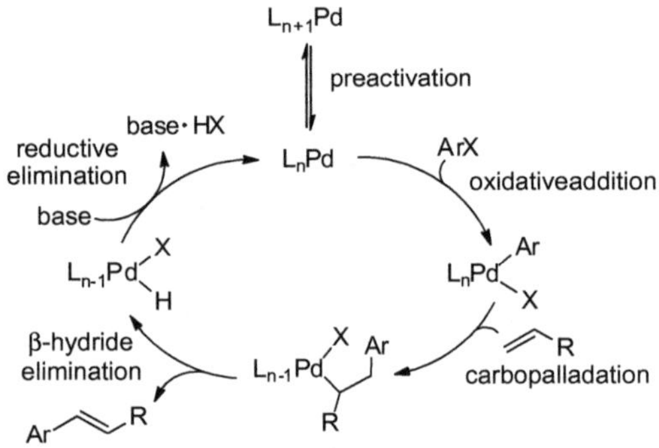

Scheme 1.18 The Heck reaction

1.7 The Heck Reaction

The palladium catalyzed direct olefination of vinylic and aromatic systems, better known as the Heck reaction, represents one of the most powerful ways for C—C bond construction. The mild reaction conditions typically applied allow a wide range of functional groups to be present on the participating coupling reagents. The Heck reaction has been considered the "*sharpening stone of palladium catalysis*" and a vast number of publications elucidating its mechanisms and applications have been reported [55]. New variations based on the early discoveries made by Richard F. Heck in the late 1960s are frequently reported, further expanding the scope and demonstrating the huge interest of this newly Nobel Prize decorated reaction.

The elementary steps comprising the overall catalytic cycle deviate from the cross coupling reactions in a number of ways as described previously (Scheme 1.18). The first step involves an oxidative addition of an aryl or vinyl halide or sulfonate to a palladium(0) complex. Next, coordination of an olefin followed by carbopalladation, dictating the regioselectivity of the reaction, results in an alkylpalladium(II) intermediate. This alkylpalladium(II) intermediate then undergoes β-hydride elimination generating the desired product in combination with a palladium(II) hydride complex. The stereochemistry of the resulting substituted double bond is determined during the β-hydride elimination. Base induced reductive elimination regenerates the palladium(0) catalyst [2].

In the following a more detailed description of the regiochemical and asymmetric aspects exhibited by the Heck reaction will be presented.

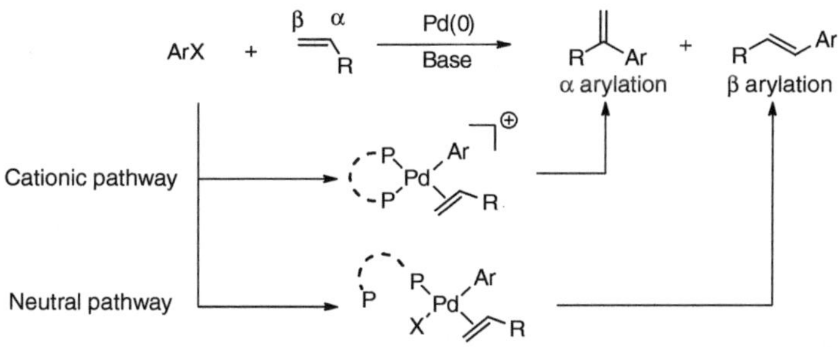

Scheme 1.19 Neutral and cationic pathways in the heck reaction

1.7.1 Regioselective Heck Reaction

During the carbopalladation two possible alkylpalladium(II) intermediates can be formed resulting in two different products; 1,1- or 1,2-disubstitued olefins. A number of strategies have been disclosed regarding the control of the regioselectivity in the Heck reaction, which can be influenced by factors such as the ligands, solvents, additives and the nature of the electrophile and the olefin. The difference between the two possible outcomes originates from the fundamental mechanisms of the carbopalladation, represented by the neutral and the cationic pathway (Schemes 1.11 and 1.19) [73, 74].

In the neutral pathway, one of the phosphines dissociates from the four-coordinated square planar palladium(II) complex, generating a neutral palladium(II) halide species, which leads to the linear β-arylated product. On the other hand, during the cationic pathway a halide or pseudohalide is replaced with the incoming olefin forming a cationic palladium(II) intermediate, which in turn leads primarily to the branched α-arylated product [75].

Nevertheless, mixtures of both isomers are often encountered implying that both pathways can be operating during the reaction. Therefore, discrimination of the two operating pathway favoring one of the isomers is of utmost importance.

In general, when a cationic pathway is desired palladium catalysts based on bidentate ligands are usually exploited in combination with electrophiles relying on weakly coordinating leaving groups such as triflates. Furthermore, electron rich olefins and solvents capable of stabilizing the intermediary cationic palladium(II) complex are typically employed. On the other hand, the neutral pathway is favored, when monodentate phosphines are used as the ancillary ligands in the palladium catalyzed olefination of aryl and vinyl halides using electron deficient alkenes such as acrylates and acrylamides.

Perhaps the most efficient way to ensure the desired regioselectivity is by the choice of electrophile. As stated, neutral palladium(II) halide complexes are typically formed, when aryl and vinyl halides are used, resulting from a strong Pd—X bond, leading to the linear alkene products. In contrast, the poor coordinating abilities of the triflate counterparts readily facilitate the formation of the

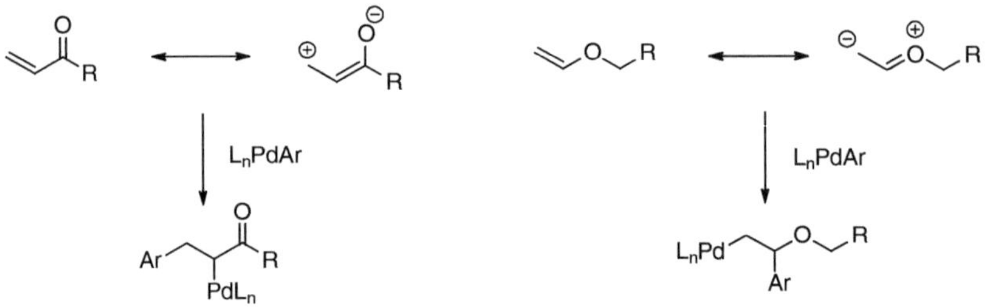

Scheme 1.20 Influence of the olefin on the regioselectivity

unsaturated cationic palladium(II) intermediate and have traditionally been applied in the regioselective α-arylation of electron rich olefins [76]. Unfortunately, the costly and toxic triflating agents in combination with instability issues displayed by the obtained aryl triflates constitute major drawbacks of this type of sulfonate. The corresponding aryl diazonium salts are also known to facilitate the intermediary cationic palladium(II) complex owing to the exceptional N_2 leaving group and weakly coordinating counterions represented by BF_4^- and PF_6^- [77].

An interchange between the two pathways can be accomplished using appropriate reaction conditions. For example, the cationic pathway can be obtained in the Heck reaction of aryl and vinyl halides by employing halide scavenging additives such as silver or thallium salts forming the cationic palladium(II) intermediate subsequent to the oxidative addition, as reported by Hallberg and coworkers [78].

In addition, the regioselectivity in the Heck reaction is considerably influenced by the electronic properties of the olefin. The difference in the regioselective outcome can be rationalized by considering the resonans structures of the two classes of alkenes (Scheme 1.20). The more electron rich carbon coordinates to the electron deficient palladium(II) center, while the aryl or vinyl group migrates to the more electron poor carbon of the olefin. However, the right combination of the olefin and the mechanistic pathway is essential in order to obtain good levels of regioselectivity. Hence, electron rich olefins are more prone to undergo α-arylation when subjected to reaction conditions favoring the cationic pathway, whereas mixtures of isomers are typically obtained if the neutral mechanism governs. On the other hand, electron deficient alkenes result primarily in β-substituted olefins regardless of the operating pathway. These trends could potentially be explained by the steric effects of the reacting species overruling the electronic properties, when the neutral pathway is in charge, whereas the electronic effects dominate during the cationic mechanism [73, 74, 79].

As an alternative to the toxic halide scavenging silver and thallium salts, polar solvents, capable of promoting the cationic palladium(II) intermediate from aryl halides, have been applied with great success producing the branched disubstituted alkenes selectively. In particular, Xiao [80] and Larhed [81] showed that aryl chlorides and bromides, usually resulting in strongly coordinated neutral palladium(II) halide complexes, could effectively be employed in the synthesis of

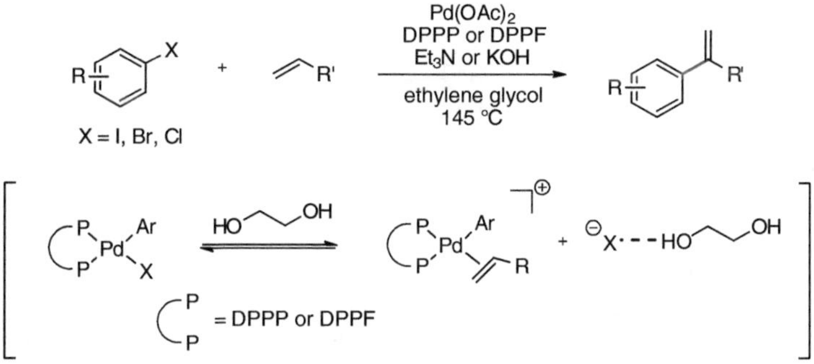

Scheme 1.21 Polar solvents in the regioselective heck reaction of aryl halides

α-substituted olefins. These findings were ascribed the stabilizing effects of high polar solvents such as ionic liquids, DMSO, DMF, H_2O or ethylene glycol on the cationic intermediate. Moreover, the hydrogen bonding abilities of the glycol enhanced the formation of the crucial cationic palladium(II) species forming the α-arylated olefins efficiently (Scheme 1.21) [82].

1.7.2 Asymmetric Heck Reaction

Among the named palladium catalyzed reactions the Heck reaction is the only one containing an inherent asymmetric aspect, which was first reported by Shibasaki and Overman, independently [83, 84]. This aspect results from the underlying mechanism of the olefination of aromatic compounds and leaves the Heck reaction more versatile and complex than the other named cross coupling reactions. As described in Sect. 1.5, certain restrictions are needed in order to accommodate the central β-hydride elimination. Hence, a *syn* coplanar geometry of the hydrogen and the palladium nucleus containing a vacant site is required resulting in the formation of a chiral center in the intermolecular Heck reaction of cyclic olefins and aryl or vinyl electrophiles (Scheme 1.22) [85]. In cases of a long-lived palladium(II) hydride complex, readdition of the palladium hydride complex and subsequent β-hydride elimination may result in isomerization of the double bond, forming the vinyl ether product [86]. The enantiomeric excess can be optimized by applying chiral ligands such as BINAP or phosphinooxazolines introduced by the group of Pfaltz [87, 88].

The intramolecular asymmetric Heck reaction has been extensively exploited in the total synthesis of a wide range of natural products forming the cyclic products attaining high yields and enantioselectivities (Scheme 1.23) [85]. The enantioselective Heck reactions are usually conducted using a combination of chiral bidentate ligands and aryl triflates, as a vacant site has to be available on the

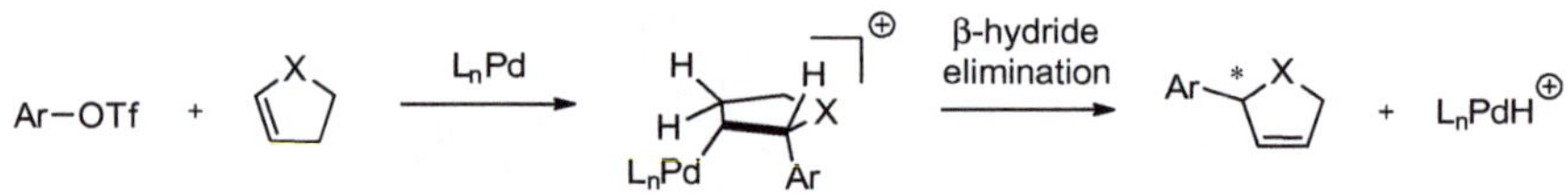

Scheme 1.22 Asymmetric heck reaction

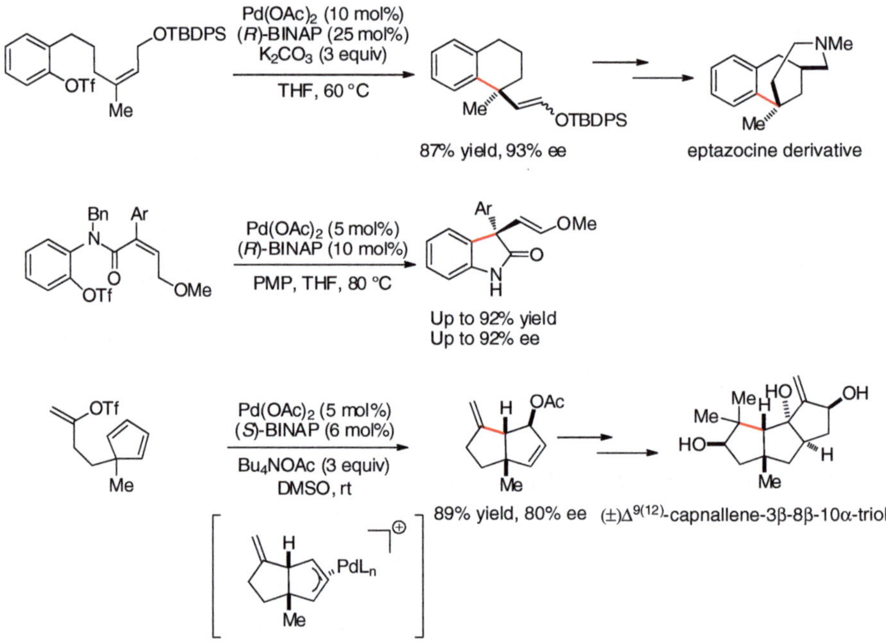

Scheme 1.23 Enantioselective heck reaction in total synthesis

central palladium after the carbopalladation, in order to facilitate the β-hydride elimination producing the resulting cyclic alkene. Alternatively, halide scavengers can be employed in the cyclizations of aryl iodides [85].

1.7.3 Applications and Variations

Due to the high levels of selectivity, functional group tolerance and the mild reaction conditions typically applied, the Heck reaction has been adapted in a number of total syntheses, some of them depicted in Scheme 1.23. In addition, industrial processes relying on the aspects displayed by the Heck reaction, have been established and resulted in large scale productions of pharmaceutical and biological active compounds (Scheme 1.24) [77, 89, 90]. Because of the high

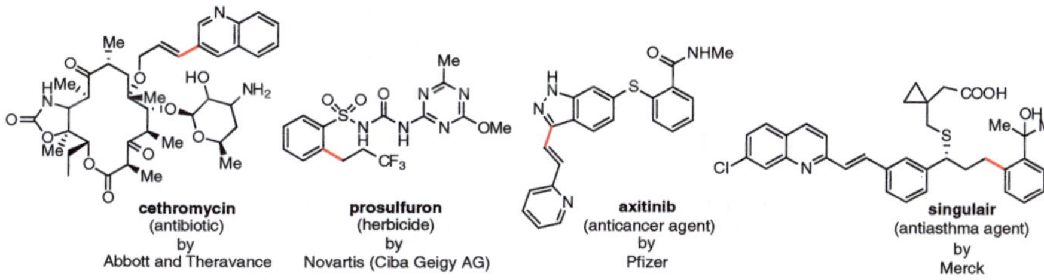

Scheme 1.24 Bioactive compounds relying on heck based strategies

flexibility and broad applications new variations and modifications of the Heck reaction are reported on a regular basis, some of which will be briefly presented in the following.

1.7.3.1 Oxidative Heck Reaction

The oxidative addition is often considered the rate determining step of the palladium catalyzed reactions and may hamper the catalytic turn-over significantly. In particular, aryl chlorides and tosylates require catalysts based on highly electron donating ancillary ligands such as trialkyl phosphines [91]. Furthermore, an inert atmosphere is typically mandatory in order to avoid the oxidation of such phosphine ligands resulting in catalyst depletion.

This, however, was circumvented by the groups of Larhed and Jung replacing the initial oxidative addition step by a transmetallation of an organometallic species to a palladium(II) complex, obtaining the equivalent arylpalladium(II) intermediate (Scheme 1.25) [92, 93]. This complex then coordinates an olefin similar to the original Heck reaction followed by a carbopalladation generating the intermediary alkylpalladium(II) complex. The final product is liberated by a β-hydride elimination producing a palladium(II) hydride, which successively reacts with molecular oxygen regenerating the palladium(II) catalyst (Scheme 1.25). In contrast to the Heck reaction, base is not necessary in order to complete the catalytic cycle, as reported by Jung and coworkers. Obviously, phosphine ligands prone to undergo oxidation resulting in weakly coordinating phosphine oxides are considered inefficient. Hence, *N,N*-bidentate ligands are most often employed in the oxidative Heck reaction, and especially phenanthroline derived ligands have proved particular adaptable (Scheme 1.26). Evidence for a palladium(II) catalytic cycle was obtained from the competition experiment of 4-iodophenyl boronic acid and *tert*-butylacrylate depicted in Scheme 1.26. Perfect chemoselectivity was achieved yielding the *tert*-butyl-3-(4-iodophenyl)acrylate exclusively leaving the iodide untouched [93]. This in turn allows for the orthogonal functionlization of aromatic systems containing multiple reactive sites.

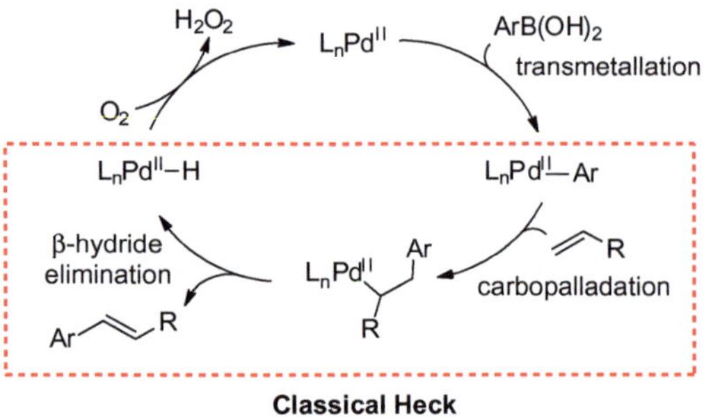

Scheme 1.25 Oxidative heck reaction

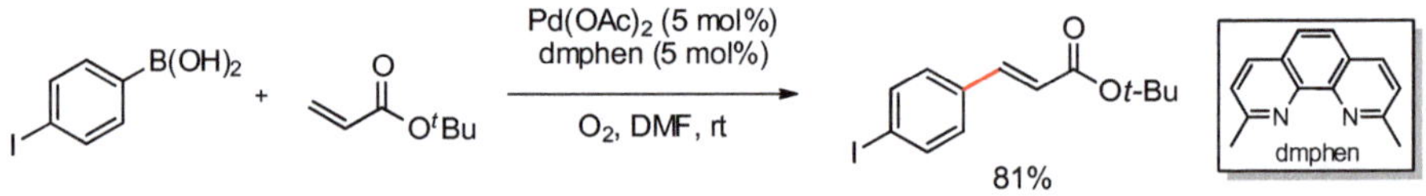

Scheme 1.26 Chemoselectivity in oxidative heck reaction

1.7.3.2 Multicomponent Heck Reaction

During the last decades efforts have been made to make transition metal catalysis even more diverse by the implementation of multiple components in a single reaction creating several chemical bonds in a one-step procedure [94].

Greany and coworkers reported on the three-component Heck reaction producing functionalized biaryl systems in one step via in situ formation of benzyne derivatives (Scheme 1.27). Mild reaction conditions were applied, and good yields were generally secured. Furthermore, the three-component route was favored over the direct Heck reaction of the aryl iodide and the olefin demonstrating the lower coordinating abilities of the electron deficient acrylates and acrylamides compared to the benzyne derivatives. The mechanism could potentially involve the oxidative addition of the aryl iodide to the starting palladium(0) complex followed by a carbopalladation to the benzyne generated in situ. The resulting biarylpalladium(II) intermediate would then react with the olefin by a Heck mechanism forming the final product (Scheme 1.27) [95].

Very recently, the group of Sigman disclosed the palladium catalyzed three-component reaction of alkenyl triflates with terminal alkenes and boronic acids (Scheme 1.27). The products resulting from a 1,1-alkene difunctionalization were obtained as the sole regioisomers and good to excellent yields were attained. The high regioselectivities were attributed to a central allylpalladium(II) complex generated from a Heck-based sequence. The intermediary allylpalladium(II) intermediate was then proposed to undergo transmetallation with the boronic acid, forming the desired benzylic olefin. These findings indicate the carbopalladation to be faster than the direct transmetallation of the boronic acid [96].

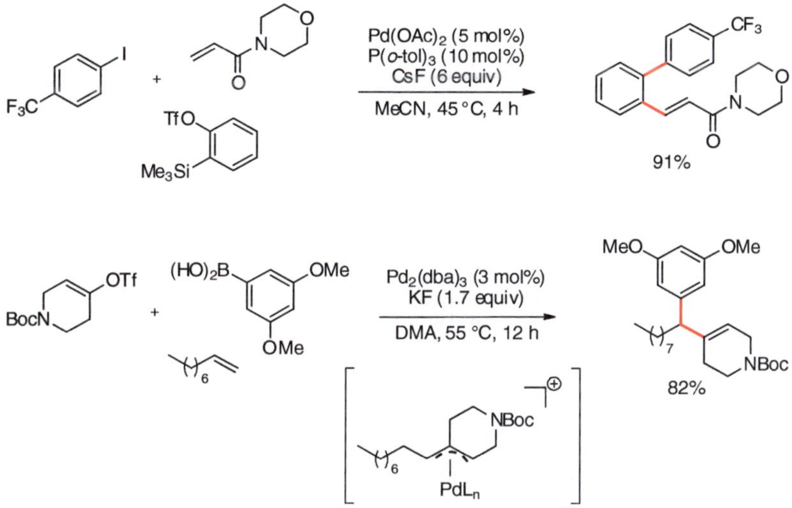

Scheme 1.27 Multicomponent couplings involving the heck reaction

1.7.3.3 Carbonylative Heck Reaction

Palladium catalyzed carbonylation reactions have received much attention during the last decades allowing the elongation of a molecule by a one carbon fragment in combination with the introduction of the highly versatile carbonyl motif using mild reaction conditions. Numerous protocols have been developed including the carbonylative Suzuki, Sonogashira and Stille reactions resulting in the formation of various ketone derived products [30]. Furthermore, the amino- and alkoxy-carbonylations generating amides and esters respectively, have been extensively investigated [30].

In contrast, the carbonylative Heck reaction, representing a variant of the palladium based multicomponent olefination reactions, has only been described to a limited extent. Nevertheless, the resulting α,β-unsaturated ketones serve as very important building blocks, as demonstrated by the many applications of the Michael addition, and constitute an attractive entry point to the assembly of heterocyclic compounds including pyridines, pyrimidines and pyrazoles etc [97, 98].

Miura and coworkers reported the first palladium catalyzed intermolecular carbonylative Heck reaction of aryl iodides and five membered olefins in the presence of carbon monoxide (Scheme 1.28) [99]. The expected 2,5-dihydrofuran product was not observed in the carbonylative coupling of 4-iodoanisole and 2,3-dihydrofuran as depicted in Scheme 1.28. Instead, isomerization of the double bond via reinsertion of the palladium(II) hydride followed by another β-hydride elimination formed the vinyl ether exclusively.

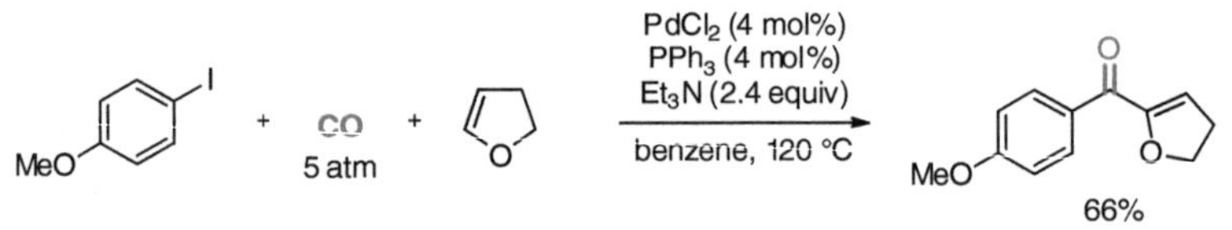

Scheme 1.28 Intermolecular carbonylative heck reaction

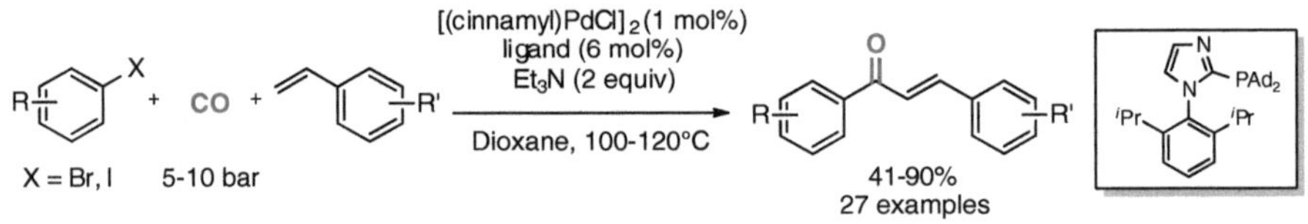

Scheme 1.29 Chalcone synthesis from aryl halides

The protocol presented by Miura was later on extended by the group of Beller, introducing styrenes as viable coupling partners to aryl halides and triflates in the synthesis of the biological relevant chalcone derivatives (Scheme 1.29) [97, 100]. Relatively high pressures of CO (5–10 bar) in combination with a catalytic system relying on a special designed imidazolylphosphine ligand proved mandatory in order to ensure acceptable turnovers of the aryl halides. On the other hand, aryl triflates could be successfully functionalized by a more conventional catalyst based on the DPPP bisphosphine ligand (Scheme 1.30).

Carbon monoxide displaying dual binding properties, acting both as a σ-donor and π-acceptor, is most likely incorporated after the oxidative addition producing an acylpalladium(II) intermediate, which is subsequently transformed into the final cholcone via carbopalladation to the styrene followed by β-hydride elimination. This was indicated by the reaction of the imidazolyl phosphine ligated *p*-tolylpalladium(II) iodide complex with carbon monoxide, generating the corresponding benzoyl palladium(II) complex (Scheme 1.31) [100].

1.8 β-Hydride Elimination Dependent Reactions

As stated in Sect. 1.5 the β-hydride elimination has been exploited in a number of processes typically originating from alkylpalladium(II) complexes forming an alkene and a palladium(II) hydride complex. In the following, some applications discovered by Skrydstrup and coworkers are presented including the palladium 1,2-migration relying on the unusual alkenyl β-hydride elimination.

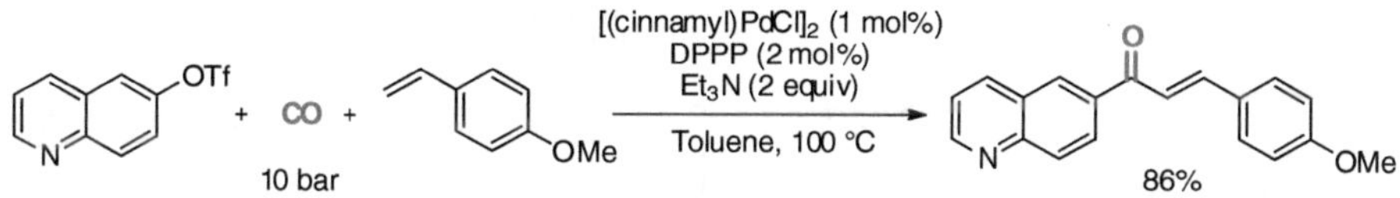

Scheme 1.30 Chalcone synthesis from aryl triflates

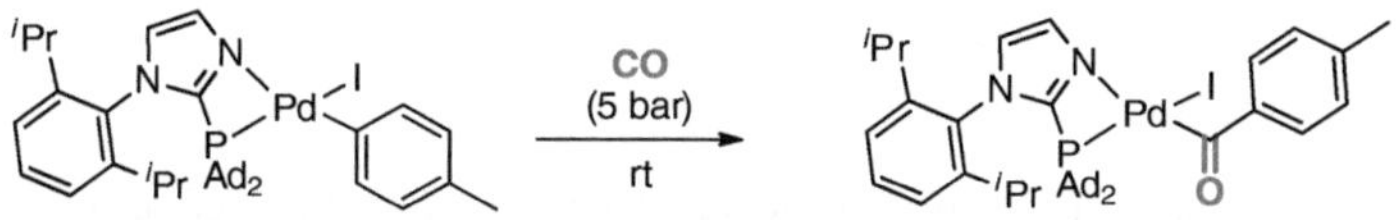

Scheme 1.31 Incorporation of carbon monoxide

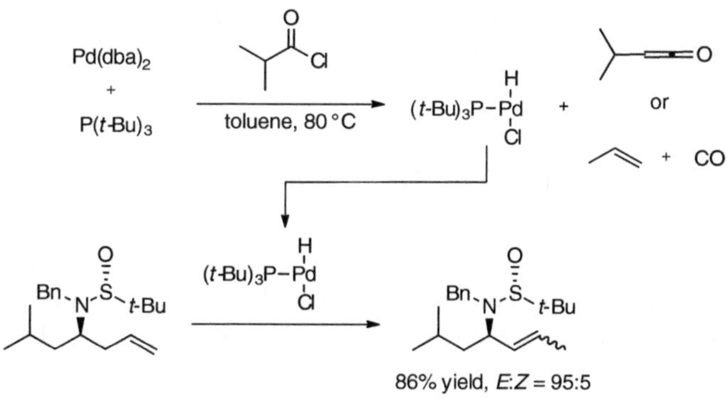

Scheme 1.32 Migration of double bond by palladium(II) hydride

1.8.1 Isomerization and Migration of Double Bonds

Recently, the group of Skrydstrup reported on the double bond isomerization or migration facilitated by an in situ generated palladium(II) hydride complex [101]. The active palladium(II) hydride complex was obtained by the oxidative addition of isobutyryl chloride to a palladium(0) species followed by either a β-hydride elimination forming a ketene or by a decarbonylation and subsequent β-hydride elimination releasing propene and carbon monoxide (Scheme 1.32). Coordination of an olefin to the catalytic active tricoordinated T-shaped palladium hydride intermediate, preceding a hydropalladation, followed by another β-hydride elimination would then produce the thermodynamically more stable alkene. This protocol was successfully expanded to include the effective isomerization of *cis*-double bonds to the corresponding *trans*-isomers. Similarly, the successive addition of the palladium(II) hydride to the double bond, rotation and β-hydride elimination formed the more stable *trans*-alkene. ^{31}P-NMR studies indicated, that the resting state of the catalytic cycle consisted of the square planar

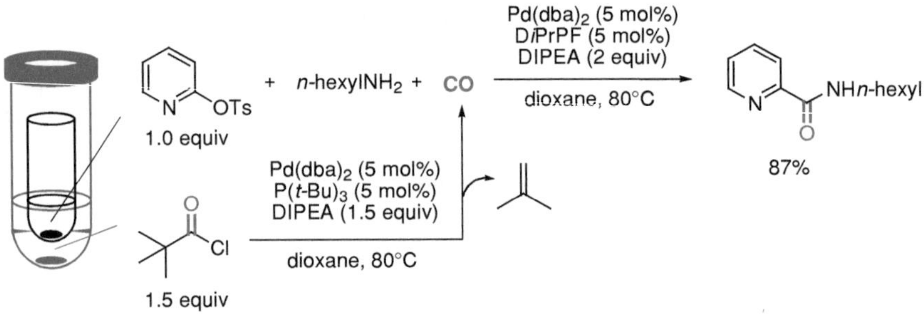

Scheme 1.33 Generation of carbon monoxide based on β-hydride elimination. Reprinted (adapted) with permission from *J. Am. Chem.Soc.* **2011**, *133*, 6061. Copyright 2011 American Chemical Society

bisphosphinechloropalladium hydride complex, isolated and characterized by Fu and coworkers, acting as a reservoir for the active catalyst [72].

1.8.2 Generation of Carbon Monoxide

Further investigations on the palladium(II) hydride generation revealed, that subjecting an acid chloride holding a tertiary alkyl group α to the carbonyl unit resulted in an effective production of carbon monoxide [102]. Thus, applying pivaloyl chloride to a chamber containing a catalytic mixture of $Pd(dba)_2$ and $P(t\text{-}Bu)_3$ in combination with stoichiometric amounts of a tertiary amine base formed carbon monoxide and isobutene efficiently. Connecting a parallel chamber, containing a mixture of 2-pyridyl tosylate and *n*-hexylamine in the presence of a palladium catalyst relying on D*i*PrPF and an amine base, to this CO generating reaction ensured the effective incorporation of the ex situ generated carbon monoxide forming the aminocarbonylated pyridine successfully (Scheme 1.33). The formation of isobutene could potentially lead to undesired byproducts such as adducts arising from hydroamination in the carbon monoxide consuming chamber depicted in Scheme 1.33. In order to address these problems, a non-volatile and crystalline carbon monoxide precursor was developed retaining the alkene byproduct in the CO generating chamber and thus avoiding cross contaminations (Scheme 1.34). Again, a hydrogen containing tertiary substituent α to the carbonyl proved essential in order to facilitate the vital β-hydride elimination subsequent to the decarbonylation step releasing the palladium(II) hydride. A variety of pharmaceutical relevant compounds and advanced intermediates were successfully synthesized in good yields using the new carbon monoxide precursor, synthesized from 9-fluorenone. Based on the same approach, the ^{13}C labeled compounds could be obtained in similar yields by applying the ^{13}C labeled CO precursor representing an unprecedented technique for carbon isotope labeling, important in the development of new drug candidates.

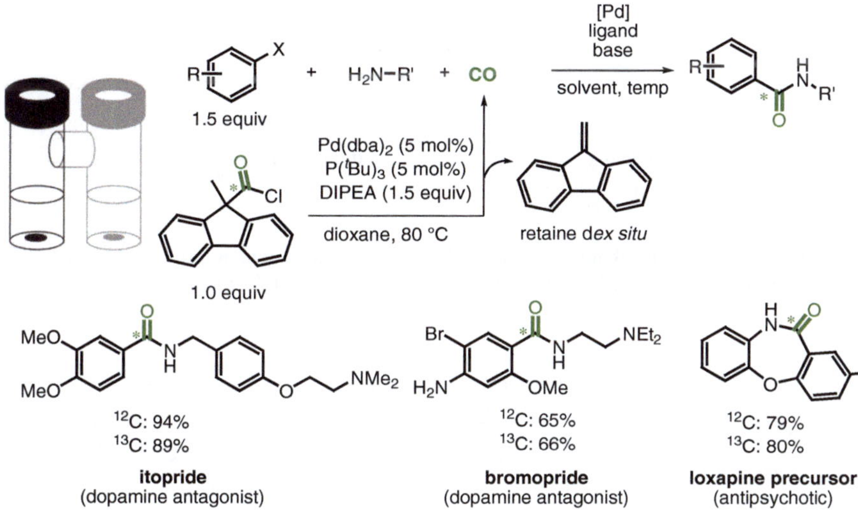

Scheme 1.34 Aminocarbonylation using ^{12}C and ^{13}C carbon monoxide

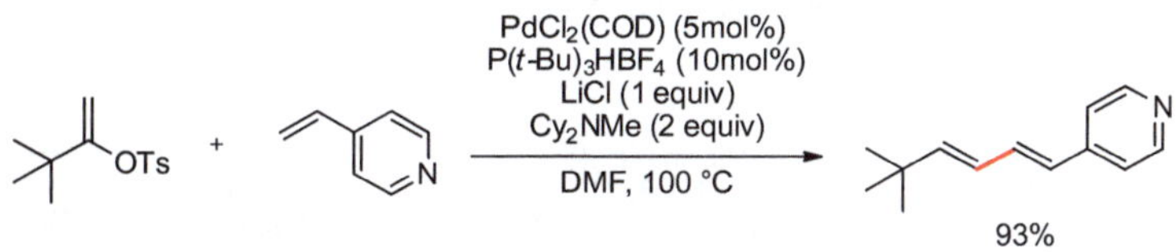

Scheme 1.35 Palladium 1,2-migration in the heck reaction

1.8.3 Palladium 1,2-Migration

During the investigations by Skrydstrup and coworkers on adopting alkenyl tosylates and phosphates as suitable electrophiles in the Heck reaction of electron deficient olefins, unexpected products originating from an apparent palladium 1,2-migration were observed. Olefination of alkenyl tosylates bearing a bulky substituent in the C1 position, catalyzed by a palladium complex relying on the sterically encumbered P(t-Bu)$_3$ in combination with LiCl and the tertiary amine base Cy$_2$NMe, efficiently produced the isomerized 1,3-dienes (Scheme 1.35) [103]. As previously described, an unsaturated tricoordinated T-shaped palladium(II) complex is formed after the oxidative addition when the sterical encumbered P(t-Bu)$_3$ is applied. Furthermore, the tosylate is most likely displaced by the stronger coordinating chloride anion resulting in a palladium(II) chloride complex analogous to the complex reported by Hartwig and coworkers [36, 44]. The increased steric effects around the palladium nucleus, induced by the bulky C1 substituent and the phosphine ligand, combined with the presence of an inherent vacant site could potentially account for the unusual alkenyl β-hydride elimination.

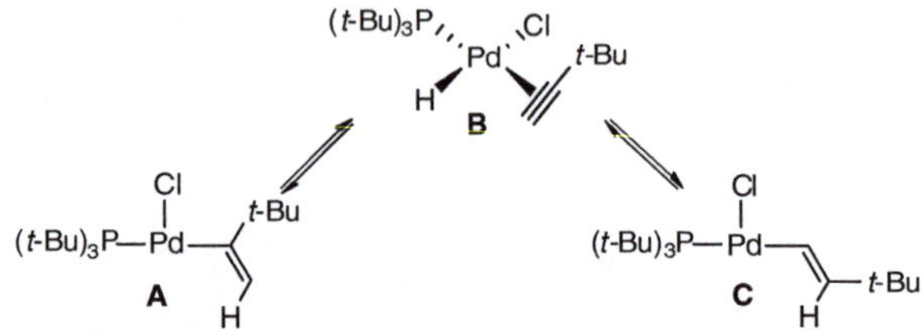

Scheme 1.36 Plausible mechanism of the palladium 1,2-migration. Reprinted (adapted) with permission from *J. Am. Chem. Soc.* **2007**, *129*, 6931. Copyright 2007 American Chemical Society

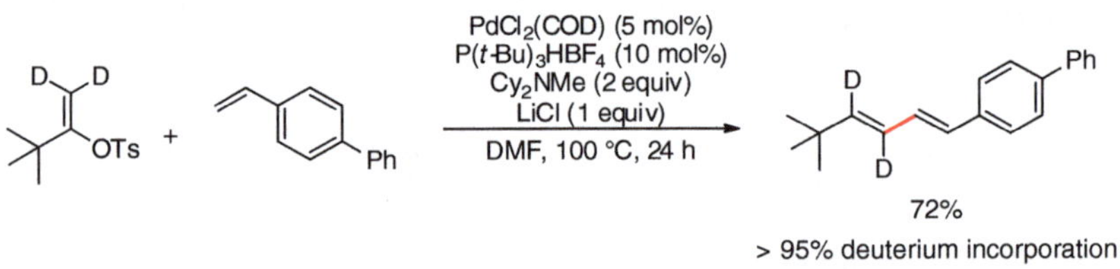

Scheme 1.37 Deuterium labeling experiment. Reprinted (adapted) with permission from *J. Am. Chem. Soc.* **2007**, *129*, 6931. Copyright 2007 American Chemical Society

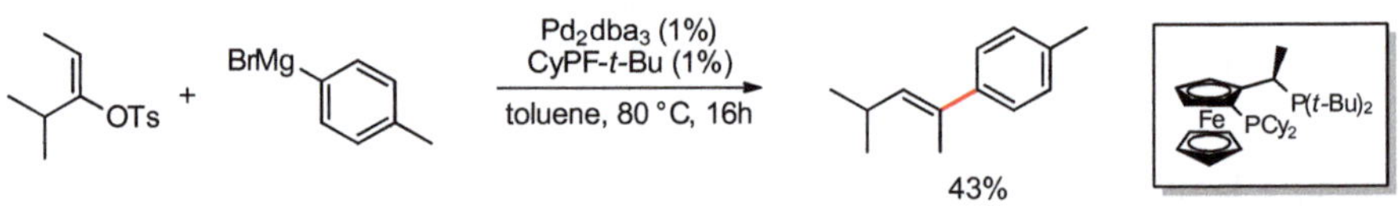

Scheme 1.38 Palladium 1,2-migration in the Kumada coupling

This would in turn generate a short-lived alkyne coordinated palladium(II) hydride intermediate. Rotation of the alkyne followed by hydropalladation producing the less hindered alkenylpalladium(II) isomer, would then ultimately lead to the isomerized diene products (Scheme 1.36). This hypothesis was supported by deuterium labeling experiments resulting in full deuterium incorporation, when the dideuterated *tert*-butylvinyl tosylate was coupled to 4-phenylstyrene (Scheme 1.37). Furthermore, DFT calculations suggested this process to be feasible under the applied reaction conditions. The pathway involving the alkenyl β-hydride elimination was revealed more favorable than the direct coordination and migratory insertion of the non-migrated alkenylpalladium(II) complex to the olefin.

The palladium 1,2-migration has only been described to a limited extent in the literature. Hartwig and coworkers reported on a single example in the Kumada coupling of alkenyl tosylates, utilizing a palladium catalyst derived from the Josiphos ligands affording the isomerized alkenyl product in a moderate 43 % isolated yield (Scheme 1.38) [104].

The increasingly common palladium migrations, activating carbon centers positioned 3, 4, 5 and 6 relative to the operative palladium center, have been described and contribute significantly to the overall applications of palladium catalysis [105]. In the following chapter the scope and limitations of the 1,2-migration in the Heck reaction will be presented illustrating the powers of yet another migratory process.

References

1. Ullmann F, Bielecki J (1901) Chem Ber 34:2174
2. Hartwig J (2010) Organotransition metal chemistry: from bonding to catalysis. University Science Books, Sausalito
3. Nicolau KC, Sorensen EJ (2003) Classics in total synthesis. Wiley-VCH, Weinheim
4. de Meijere A, Diedrich F (2004) Metal-catalyzed cross-coupling reactions, 2nd edn. Wiley-VCH, Weinheim
5. Amatore C, Carré E, Jutand A, M'Barki MA, Meyer G (1995) Organometallics 14:5605
6. Trceziak AM, Ziolkowski J (2005) J Coord Chem Rev 249:2308
7. Fors BP, Krattiger P, Strieter E, Buchwald SL (2008) Org Lett 10:3505
8. Amatore C, Broeker G, Jutand A, Khalil F (1997) J Am Chem Soc 119:5176
9. Amatore C, Carré E, Jutand A, Medjour Y (2002) Organometallics 21:4540
10. Amatore C, Jutand A (1999) J Organomet Chem 576:254
11. Vo GD, Hartwig JF (2009) J Am Chem Soc 131:11049
12. Ogata T, Hartwig JF (2008) J Am Chem Soc 130:13848
13. Littke AF, Schwarz L, Fu GC (2002) J Am Chem Soc 124:6343
14. Biscoe MR, Fors BP, Buchwald SL (2008) J Am Chem Soc 130:6686
15. Kinzel T, Zhang Y, Buchwald SL (2010) J Am Chem Soc 132:14073
16. Stambuli JP, Kuwano R, Hartwig JF (2002) Angew Chem Int Ed 41:4746
17. Liu W, Chen D, Zhu X–Z, Wan X–L, Hou X–L (2009) J Am Chem Soc 131:8734
18. Marion N, Navarro O, Mei J, Stevens ED, Scott NM, Nolan SP (2006) J Am Chem Soc 128:4101
19. Tolman CA (1977) Chem Rev 77:313
20. Birkholz (née Gensow) M–N, Freixa Z, van Leeuwen PWNM (2009) Chem Soc Rev 38:1099
21. Stambuli JP, Bühl M, Hartwig JF (2002) J Am Chem Soc 124:9346
22. Hartwig JF (2008) Acc Chem Res 41:1534
23. Hamann BC, Hartwig JF (1998) J Am Chem Soc 120:3694
24. Martinelli JR, Watson DA, Freckmann DMM, Barder TE, Buchwald SL (2008) J Org Chem 73:7102
25. Fu GC (2008) Acc Chem Res 41:1555
26. Netherton MR, Fu GC (2001) Org Lett 3:4295
27. Martin R, Buchwald SL (2008) Acc Chem Res 41:1461
28. Fors BP, Watson DA, Biscoe MR, Buchwald SL (2008) J Am Chem Soc 130:13552
29. Shen Q, Ogata T, Hartwig JF (2008) J Am Chem Soc 130:6586
30. Brennführer A, Neumann H, Beller M (2009) Angew Chem Int Ed 48:4114
31. Alcazar-Roman LM, Hartwig JF (2002) Organometallics 21:491
32. Newman SG, Lautens M (2010) J Am Chem Soc 132:11416
33. Senn HM, Ziegler T (2004) Organometallics 23:2980
34. Netherton MR, Fu GC (2002) Angew Chem Int Ed 41:3910
35. Hartwig JF, Paul F (1995) J Am Chem Soc 117:5373

36. Barrios-Landeros F, Carrow BP, Hartwig JF (2009) J Am Chem Soc 131:8141
37. Amatore C, Jutand A (2000) Acc Chem Res 33:314
38. Roy AH, Hartwig JF (2004) Organometallics 23:194
39. Amatore C, Azzabi M, Jutand A (1991) J Am Chem Soc 113:8375
40. Littke AF, Dai C, Fu GC (2000) J Am Chem Soc 122:4020
41. Alcazar-Roman LM, Hartwig JF, Rheingold AL, Liable-Sands LM, Guzei IA (2000) J Am Chem Soc 122:4618
42. Dierkes P, van Leeuwen PWNM (1999) J Chem Soc Dalton Trans 1519
43. Freixa Z, van Leeuwen PWNM (2003) J Chem Soc Dalton Trans 1890
44. Roy AH, Hartwig JF (2003) J Am Chem Soc 125:8704
45. Tang Z–Y, Hu Q–S (2004) J Am Chem Soc 126:3058
46. Tang Z–Y, Hu Q–S (2006) J Org Chem 71:2167
47. Rosen BM, Quasdorf KW, Wilson DA, Zhang N, Resmerita A-M, Garg NK, Percec V (2011) Chem Rev 111:1346
48. Casado AL, Espinet P (1998) J Am Chem Soc 120:8978
49. Corbet J–P, Mignani G (2006) Chem Rev 106:2651
50. Littke AF, Fu GC (2002) Angew Chem Int Ed 41:4176
51. Labadie JW, Stille JK (1983) J Am Chem Soc 105:6129
52. Heck RF (1968) J Am Chem Soc 90:5518
53. Mizoroki T, Mori K, Ozaki A (1971) Bull Chem Soc Jpn 44:581
54. Heck RF, Nolley JP Jr (1972) J Org Chem 37:2320
55. Beletskaya IP, Cheprakov AV (2000) Chem Rev 100:3009
56. Stambuli JP, Incarvito CD, Bühl M, Hartwig JF (2004) J Am Chem Soc 126:1184
57. Nakamura A, Ito S, Nozaki K (2009) Chem Rev 109:5215
58. Trost BM, Gutierrez AC, Ferreira EM (2010) J Am Chem Soc 132:9206
59. John J, U I, Suresh E, Radhakrishnan KV (2009) J Am Chem Soc 131:5042
60. Zhou J, Fu GC (2003) J Am Chem Soc 125:14726
61. Trost BM, Gutierrez AC, Ferreira EM (2008) J Am Chem Soc 130:16176
62. Lindhardt (née Hansen) AT, Mantel MLH, Skrydstrup T (2008) Angew Chem Int Ed 47:2668
63. Fujita K, Yoshida T, Imori Y, Yamaguchi R (2011) Org Lett 13:2278
64. de Meijere A, Meyer FE (1994) Angew Chem Int Ed 33:2379
65. Anslyn EV, Dougherty DA (2006) Modern physical organic chemistry. University Science Books, Sausalito
66. Hegedus LS (1999) Transition metals in the synthesis of complex organic molecules. 2nd edn. University Science Books, Sausalito
67. Hartwig JF (1936) Inorg Chem 2007:46
68. Tatsumi K, Hoffmann R, Yamamoto A, Stille JK (1981) Bull Chem Soc Jpn 54:1857
69. Christmann U, Vilar R (2005) Angew Chem Int Ed 44:366
70. Culkin DA, Hartwig JF (2004) Organometallics 23:3398
71. Grushin VV, Marshall WJ (2006) J Am Chem Soc 128:12644
72. Hills ID, Fu GC (2004) J Am Chem Soc 126:13178
73. Liu S, Berry N, Thomson N, Pettman A, Hyder Z, Mo J, Xiao J (2006) J Org Chem 71:7467
74. Cabri W, Candiani I (1995) Acc Chem Res 28:2
75. Mo J, Xu L, Xiao J (2005) J Am Chem Soc 127:751
76. Vallin KSA, Zhang Q, Larhed M, Curran DP, Hallberg A (2003) J Org Chem 68:6639
77. Roglans A, Pla-Quintana A, Moreno-Mañas M (2006) Chem Rev 106:4622
78. Sonesson C, Larhed M, Nyqvist C, Hallberg A (1996) J Org Chem 61:4756
79. von Schenck H, Åkermark B, Svensson M (2003) J Am Chem Soc 125:3503
80. Ruan J, Xiao J (2011) Acc Chem Res 44:614
81. Arvela RK, Pasquini S, Larhed M (2007) J Org Chem 72:6390
82. Ruan J, Iggo JA, Berry NG, Xiao J (2010) J Am Chem Soc 132:16689
83. Sato Y, Sodeoka M, Shibasaki M (1989) J Org Chem 54:4738
84. Carpenter NE, Kucera DJ, Overman LE (1989) J Org Chem 54:5846

85. Dounay AB, Overman LE (2003) Chem Rev 103:2945
86. Loiseleur O, Meier P, Pfaltz A (1996) Angew Chem Int Ed 35:200
87. Loiseleur O, Hayashi M, Schmees N, Pfaltz A (1997) Synthesis 1338
88. Shibasaki M, Vogl EM (1999) J Organomet Chem 576:1
89. Magano J, Dunetz JR (2011) Chem Rev 111:2177
90. de Vries JG (2001) Can J Chem 79:1086
91. Ahlquist M, Fristrup P, Tanner D, Norrby P-O (2006) Organometallics 25:2066
92. Enquist P-A, Nilsson P, Sjöberg P, Larhed M (2006) J Org Chem 71:8779
93. Yoo KS, Yoon CH, Jung KW (2006) J Am Chem Soc 128:16384
94. Siamaki AR, Arndtsen BA (2006) J Am Chem Soc 128:6050
95. Henderson JL, Edwards AS, Greaney MF (2007) Org Lett 9:5589
96. Liao L, Jana R, Urkalan KB, Sigman MS (2011) J Am Chem Soc 133:5784
97. Wu X-F, Neumann H, Beller M (2010) Angew Chem Int Ed 49:5284
98. Wu X-F, Jiao H, Neumann H, Beller M (2011) ChemCatChem 3:726
99. Satoh T, Itaya T, Okuro K, Miura M, Nomura M (1995) J Org Chem 60:7267
100. Wu X-F, Neumann H, Spannenberg A, Schulz T, Jiao H, Beller M (2010) J Am Chem Soc 132:14596
101. Gauthier D, Lindhardt AT, Olsen EPK, Overgaard J, Skrydstrup T (2010) J Am Chem Soc 132:7998
102. Hermange P, Lindhardt AT, Taaning RH, Bjerglund K, Lupp D, Skrydstrup T (2011) J Am Chem Soc 133:6061
103. Hansen AL, Ebran J-P, Ahlquist M, Norrby P-O, Skrydstrup T (2006) Angew Chem Int Ed 45:3349
104. Limmert ME, Roy AH, Hartwig JF (2005) J Org Chem 70:9364
105. Mota AJ, Dedieu A (2006) Organometallics 25:3130

85. Dounay AB, Overman LE (2003) Chem Rev 103:2945
86. [illegible] Angew Chem Int Ed 45:[illegible]
87. [illegible] (2009) Synthesis [illegible]
88. [illegible]
89. [illegible] Chem Rev [illegible]
90. de [illegible] Chem [illegible]
91. [illegible] (2000) Organomet [illegible]
92. [illegible] J Org Chem [illegible]
93. [illegible] Soc [illegible]
94. [illegible] Am Chem Soc [illegible]
95. [illegible] Chem Lett [illegible]
96. [illegible]
97. [illegible] Chem Int Ed [illegible]
98. [illegible]
99. [illegible]
100. [illegible]
101. [illegible]
102. [illegible]
103. [illegible]
104. [illegible]
105. [illegible]

Chapter 2
Palladium 1,2-Migration in the Heck Reaction

2.1 Introduction

A more thorough investigation of the palladium 1,2-migration in the Heck reaction was desired, in order to clarify the underlying factors affecting the migratory process. In particular, full control of the isomerization of the 1,3-dienes obtained from the Heck reaction of alkenyl C–O coupling partners and electron deficient olefins would be of high interest. This would in turn represent a novel approach towards the construction of linear disubstituted dienes starting from 1,1-disubstituted alkenyl electrophiles.

2.2 Background

As mentioned previously, different palladium migrations have been reported in the literature typically involving the intramolecular migration of a palladium(II) complex from one carbon to another [1]. The exact mechanism of these migrations is highly dependent on the particular migratory process and may involve palladium(II)- and/or (IV) hydride intermediates. DFT calculations provided by Dedieu and coworkers suggested, that a palladium(IV) hydride intermediate was operating in sterically hindered and geometrically constrained systems such as in the 1,3-migration. On the other hand, a palladium(II) intermediate was found to be involved in the less hindered 1,5- and 1,6-migrations, whereas the 1,4-migration consisted of a mixture of the two pathways (Scheme 2.1) [2].

During the last decades much effort has been directed towards the implementation of alcohol derived electrophiles different from the costly and often unstable triflates and nonaflates. The corresponding phosphates and tosylates, representing a more stable and less costly alternative to the traditional sulfonates, have been successfully utilized in a number of processes including the transition metal catalyzed Suzuki, Kumada and Negishi reactions reported by Skrydstrup and

T. M. Gøgsig, *New Discoveries on the β-Hydride Elimination*, Springer Theses,
DOI: 10.1007/978-3-642-32099-6_2, © Springer-Verlag Berlin Heidelberg 2012

Scheme 2.1 Palladium 1,3- to 1,6-migrations

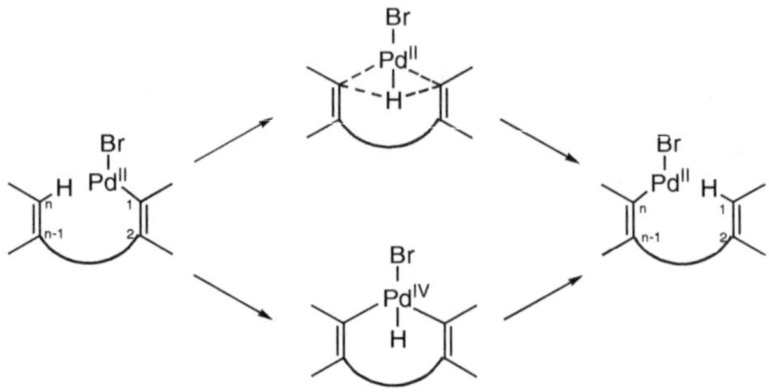

coworkers [3–6]. In contrast, aryl and alkenyl tosylates and phosphates have only been scarcely employed in the palladium catalyzed Heck reaction illustrating the distinct requirements of the Heck reaction compared to the other cross coupling reactions [7, 8].

2.3 Results and Discussion

In order to gain control of the palladium 1,2-migration, suitable reaction conditions, facilitating the non-migrated Heck products using non-activated alkenyl electrophiles, were needed (Table 2.1). The corresponding alkenyl phosphates were chosen as test substrates in the Heck reaction of *N-tert*-butylacrylamide, employing the precatalyst $PdCl_2$(COD) and the tertiary amine base Cy_2NMe in DMF at 100 °C. Preliminary screenings of the conventional bidentate ligands including DPPE, DPPP, DPPF, D*t*BPF, BINAP and the Josiphos ligands did not accommodate the desired olefination. Applying bidentate ligands would potentially impede the 1,2-migration, as a saturated square planar palladium(II) complex is formed after the oxidative addition, occupying the vacant site crucial to the alkenyl β-hydride elimination involved in the 1,2-migration. Changing to the dialkylbiaryl ligands introduced by the group of Buchwald proved more promising, and a 48 % conversion was observed employing XPhos, based on ^{1}H NMR analysis of the crude reaction mixture. This resulted in a 37 % isolated yield of the non-isomerized diene exclusively (entries 1–5) [9] These monophosphine ligands are considered bulky and electron rich and have been demonstrated to coordinate the palladium centre in bidentate fashion through a η^1 palladium interaction of the 1' ipso carbon, which could possibly explain the non-migrating properties of the catalyst [10, 11]. The palladacycle SK-CC01-A performed equally well, whereas the cataCXium ligands proved ineffective (entries 6–11) [12]. Further optimizations were performed using XPhos revealing a slightly increased activity, when DMF and dioxane were combined in a 1:1 ratio (entry 14). Changing the solvent to dioxane or elevating the temperature to 140 °C proved ineffective (entries 12 and 13). Next, the amount of added LiCl was investigated. As demonstrated in Table 2.2 improved catalytic activity was obtained, when increased amounts of LiCl were added to the

Table 2.1 Optimization of the non-migrating Heck reaction

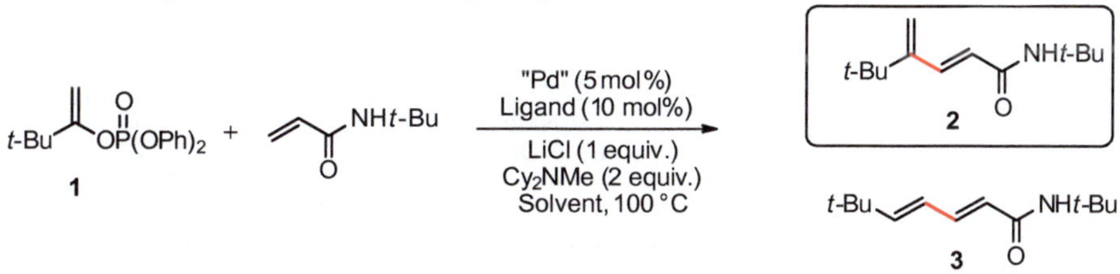

Entry	"Pd"/ligand	Solvent	Conversion[a] (yield)[b] (%)	2:3
1	$PdCl_2$(COD), S-Phos	DMF	33	>99:1
2	$PdCl_2$(COD), X-Phos	DMF	48 (37)	>99:1
3	$PdCl_2$(COD), *L1*	DMF	<5	–
4	$PdCl_2$(COD), *L2*	DMF	<5	–
5	$PdCl_2$(COD), JohnPhos	DMF	38	>99:1
6	$PdCl_2$(COD), CataCXium PtB	DMF	22	–
7	$PdCl_2$(COD), CataCXium PCy	DMF	<5	–
8	$PdCl_2$(COD), CataCXium PIntB	DMF	<5	–
9	$PdCl_2$(COD), CataCXium PInCy	DMF	<5	–
10	SK-CC01-A	DMF	47	>99:1
11	SK-CC02-A	DMF	13	>99:1
12	$PdCl_2$(COD), X-Phos[c]	DMF	21	54:46
13	$PdCl_2$(COD), X-Phos	Dioxane	23	>99:1
14	$PdCl_2$(COD), X-Phos	DMF/dioxane[d]	56 (45)	>99:1

[a] Conversion determined by ^1H NMR analysis of crude reaction mixture. [b] Isolated yields. [c] Reaction run at 140 °C. [d] 1:1 ratio

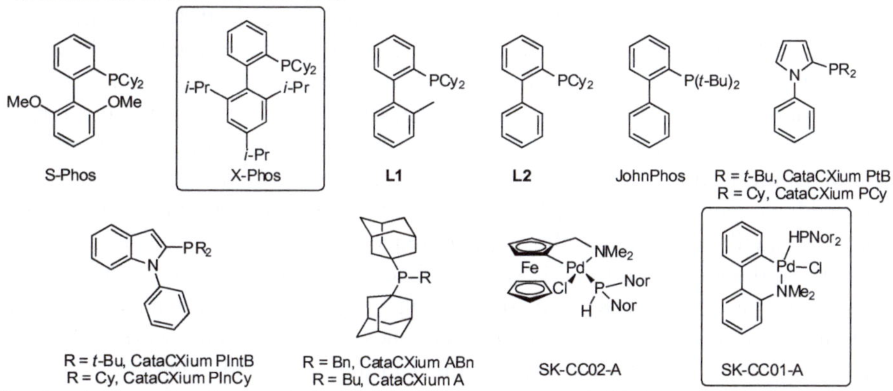

Reprinted (adapted) with permission from *J. Am. Chem. Soc.* **2007**, *129*, 6931. Copyright 2007 American Chemical Society

reaction mixture attaining full conversion with 10 equivalents added (entry 3). The most reliable system was acquired, when a catalytic loading of 5 % with a palladium:ligand ratio of 1:1 was employed, exclusively providing the non-isomerized diene in a 91 % isolated yield (entry 7). On the other hand, no product was observed, when substituting LiCl with NaCl leading to catalyst decomposition

Table 2.2 Studies on the effect of additives

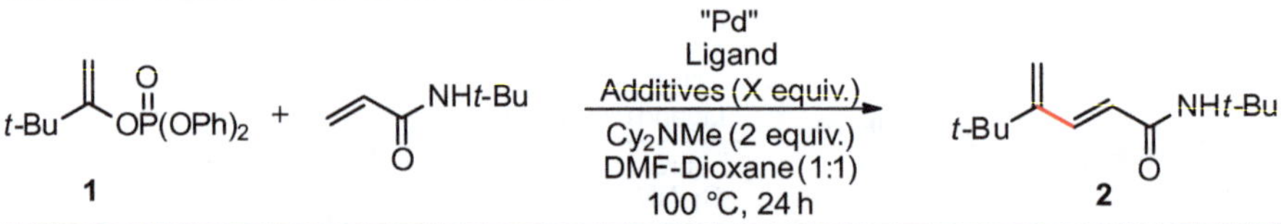

Entry	Catalytic system (%)	Additives (X equiv.)	Conversion[a] (yield)[b] (%)
1	$PdCl_2(COD)$ (2.5), X-Phos (5)	LiCl (2)	70
2	$PdCl_2(COD)$ (2.5), X-Phos (5)	LiCl (4)	86
3	$PdCl_2(COD)$ (2.5), X-Phos (5)	LiCl (10)	99
4	$PdCl_2(COD)$ (2.5)[c]	LiCl (10)	10
5	$PdCl_2(COD)$ (2.5), X-Phos (5)	NaCl (10)	<5
6	$PdCl_2(COD)$ (2.5), X-Phos (2.5)	LiCl (10)	77
7	$PdCl_2(COD)$ (5), X-Phos (5)	LiCl (10)	100 (91)
8	SK-CC01-A (5)	LiCl (10)	90

[a] Conversion determined by ^1H NMR analysis of crude product. [b] Isolated yields. [c] Only DMF was used as solvent

after a short period of time (entry 5). Once again, the palladacycle SK-CC01-A displayed similar reactivity resulting in 90 % conversion under the applied reaction conditions (entry 8). An increased concentration of the anionic palladium(0) species prior to the oxidative addition, as described by Amatore and Jutand, could potentially explain the beneficial effects of the added LiCl [13].

With the optimized reaction conditions in hand, we set forth to investigate the generality of the Heck coupling between different alkenyl phosphates and electron deficient olefins (Scheme 2.2). Alkenyl phosphates holding primary and secondary alkyl groups on the C1 position coupled successfully resulting in the non-migrated dienes exclusively. However, when styrene was introduced as the coupling partner, partial isomerisation of the π-system was observed, possibly facilitated by the intermediary palladium(II) hydride. C1-quarternary and aryl vinyl phosphates coupled equally well under the applied reaction conditions affording the non-isomerized dienes in good yields when coupled to acrylate, styrene or acrylamide illustrating the pronounced non-isomerizing properties of the catalytic system. Functional groups such as silyl ethers, lactones and carbamates were well tolerated, although a modest yield was obtained in the case of the lactone.

Finally, alkenyl phosphates bearing a substituent in the 2 position of the alkene were tested. Starting from isomeric pure 1,2-disubstituted vinyl phosphates, the Heck reaction led to severe scrambling of the stereochemistry of the double bond in the final product, demonstrating the palladium(II) hydride generated upon the β-hydride elimination to be relatively long-lived. Nevertheless, good yields of the non-migrated dienes could be secured after column chromatography.

Having established conditions for the Heck reaction employing alkenyl phosphates as electrophiles obtaining the non-migrated products, further exploration of

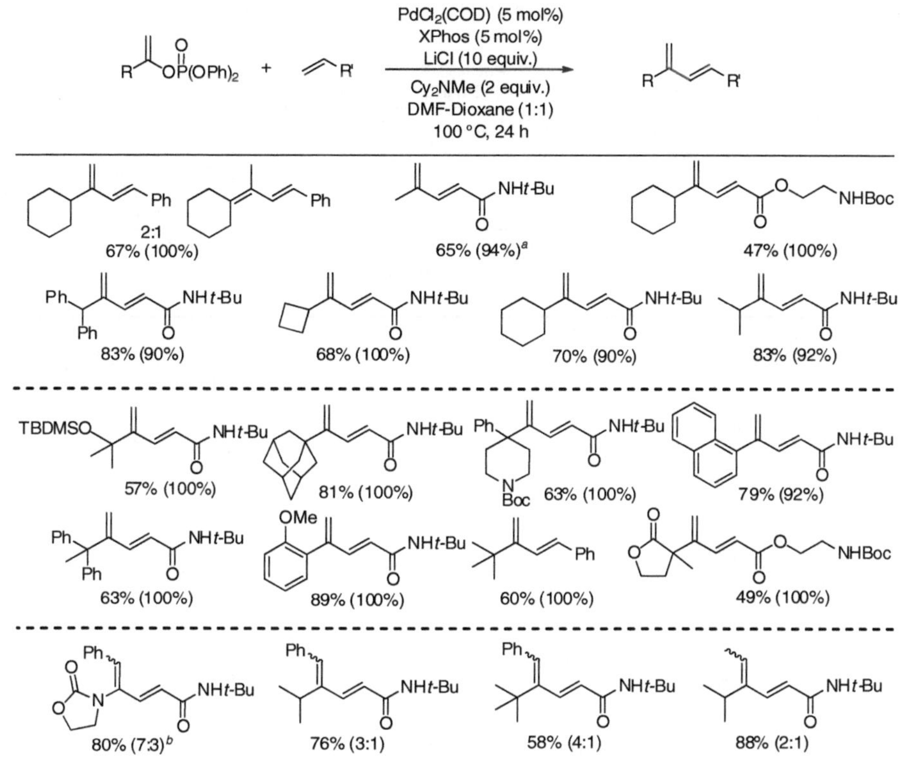

Isolated yields. Conversion in parentheses determined by ^{1}H NMR analysis of crude product. [a] Reaction after 72 h. [b] Configuration of the major isomer was determined by single-crystal X-ray analysis to be the Z-isomer.

Scheme 2.2 Heck reaction without palladium 1,2-migration

the palladium 1,2-migration in the Heck reaction was desired (Table 2.3). Preliminary investigations revealed the alkenyl phosphates holding an allylic hydrogen unreactive, and full recovery of the starting materials was attained. Furthermore, the 1,2-disubstituted vinyl phosphates depicted in Scheme 2.2 did not undergo successful coupling, suggesting that the palladium 1,2-migration is limited to substrates carrying a quaternary substituent on the C1 position. Potentially, this could be explained by an elongation of the C1–O bond, when sterically encumbered susbstituents are introduced on the C1 position resulting in a more feasible oxidative addition. Hence, attention was directed towards different C1-quaternary alkenyl phosphates, as depicted in Table 2.3. The palladium 1,2-migration proved effective for the alkenyl phosphates depicted in entries 1–4, obtaining the isomerized diene with regioselectivitites greater than 9:1. The low levels of palladium 1,2-migration, observed in the Heck reaction employing a lactone derived alkenyl phosphate, could possibly be explained by an intramolecular coordination of the carbonyl oxygen to the palladium center after the

Table 2.3 Heck couplings of C1-quaternary alkyl vinyl phosphates with 1,2-migration

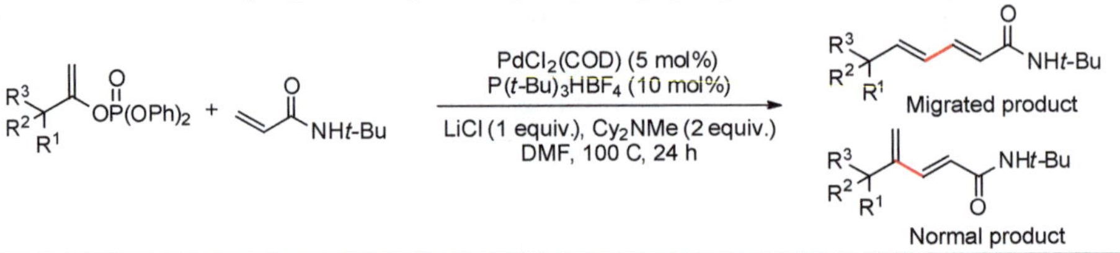

Entry	Vinyl phosphate	Conversion[a] (%)	Yield[b] (%) (migration:non-migration)[c]	Product
1	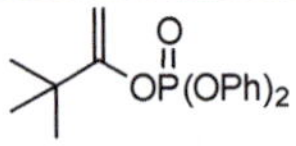	59	50 (93:7)	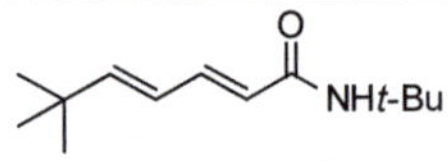
2	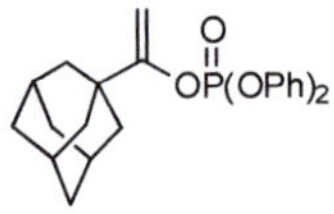	100	73 (9:1)	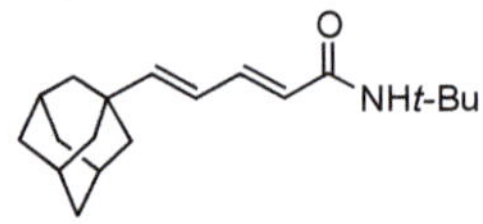
3	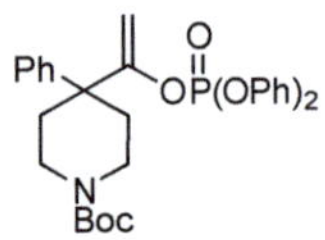	84	75 (9:1)	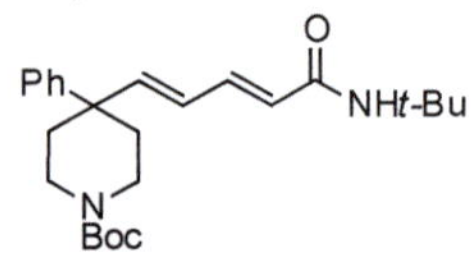
4	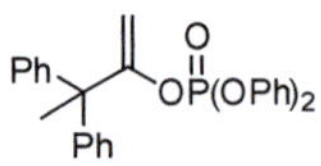	44	36 (>99:1)	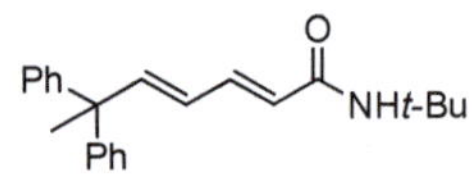
5	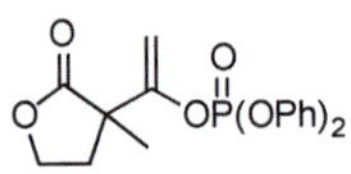	100	81 (2:5)	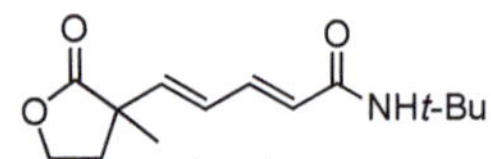 minor isomer
6	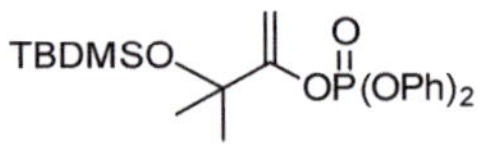	n.r.[d]	–	–
7	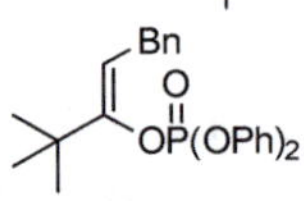	n.r.[d]	–	–

[a] Conversion determined by ^{1}H NMR analysis of crude product. [b] Isolated yields. [c] Ratio of isomers determined by ^{1}H NMR spectrum of crude product. [d] No reaction

Reprinted (adapted) with permission from *J. Am. Chem. Soc.* **2007**, *129*, 6931. Copyright 2007 American Chemical Society

oxidative addition, preventing the migration to occur (entry 5). In contrast to the Heck reaction, employing the catalytic system based on XPhos, producing the non-isomerized dienes, no reaction was observed, when the silylether derived alkenyl phosphate was subjected to the conditions promoting the 1,2-migration (entry 6). This could perhaps be explained by a deprotection of the TBDMS-group by the LiCl in DMF at elevated temperatures. Similarly, the 1,2-disubstituted alkenyl phosphates surprisingly turned out nonreactive (entry 7).

Table 2.4 Heck couplings of aryl vinyl phosphates with 1,2-migration

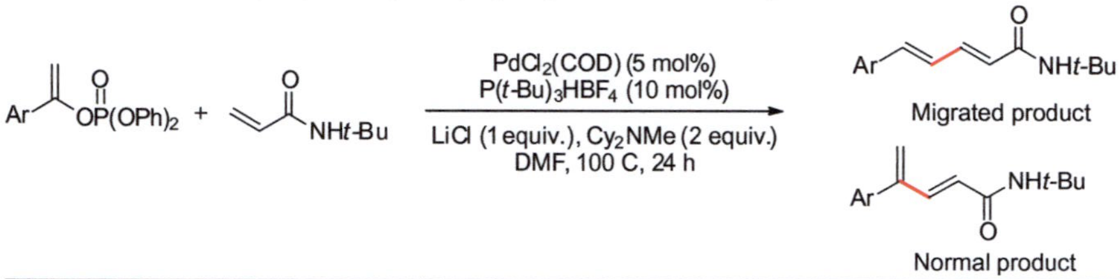

Entry	Vinyl phosphate	Conversion[a] (%)	Yield[b] (%) (migration:normal)[c]	Product
1	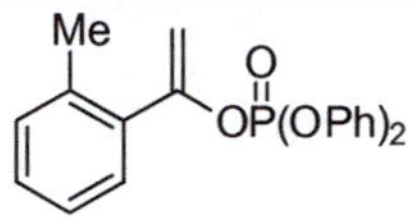	100	95 (10:1)	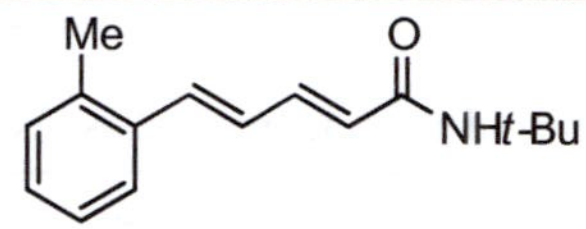
2	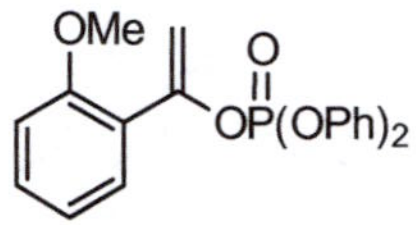	90	80 (1:1)[d]	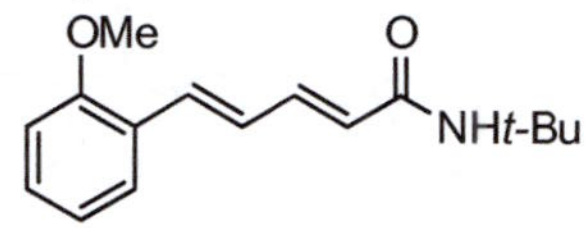
3	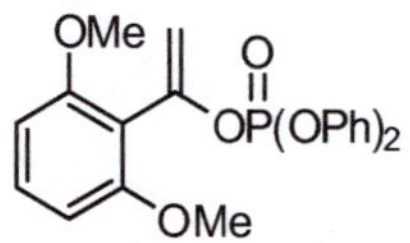	100	96 (3:1)	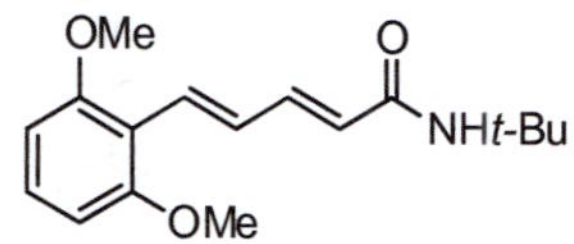
4	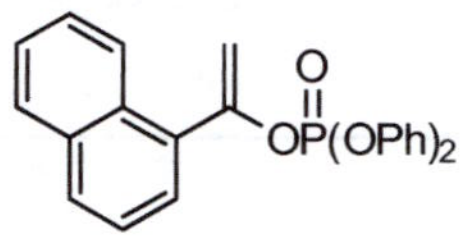	100	91 (1:2)	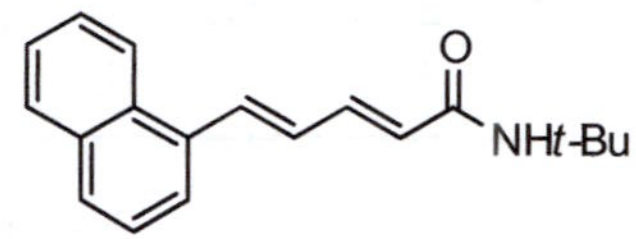

(continued)

Table 2.4 (continued)

Entry	Vinyl phosphate	Conversion[a] (%)	Yield[b] (%) (migration:normal)[c]	Product
5	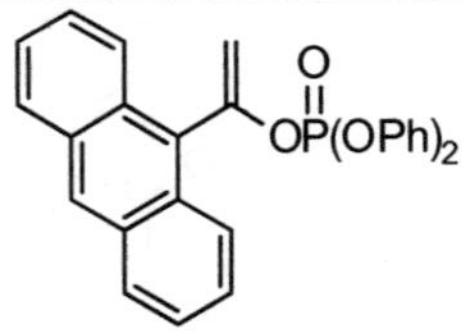	100	74 (>99:1)	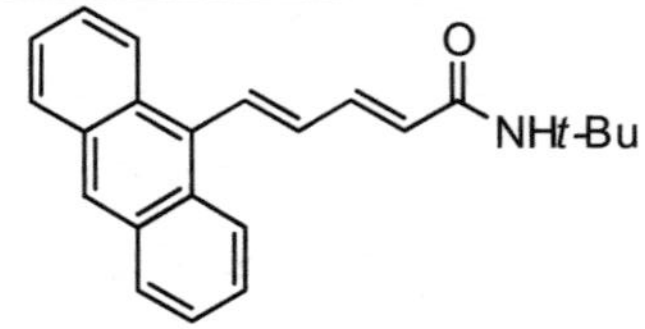
6	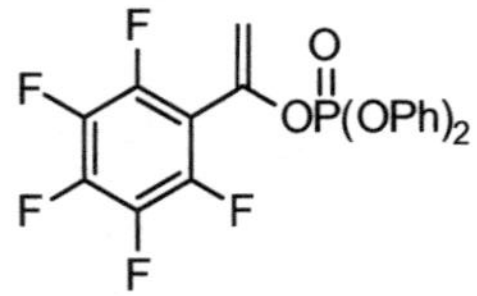	100	81 (<1:99)	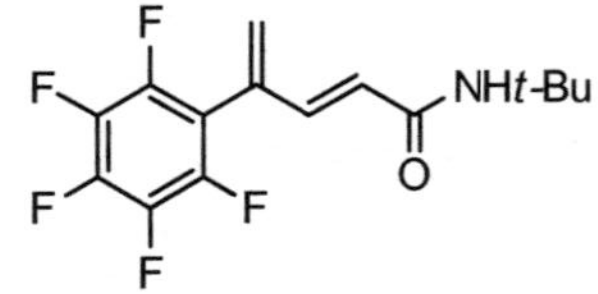

[a] Conversion determined by ^{1}H NMR spectrum of crude product. [b] Isolated yields. [c] Ratio of isomers determined by ^{1}H NMR spectrum of crude product.
[d] Reaction carried out at 60 °C

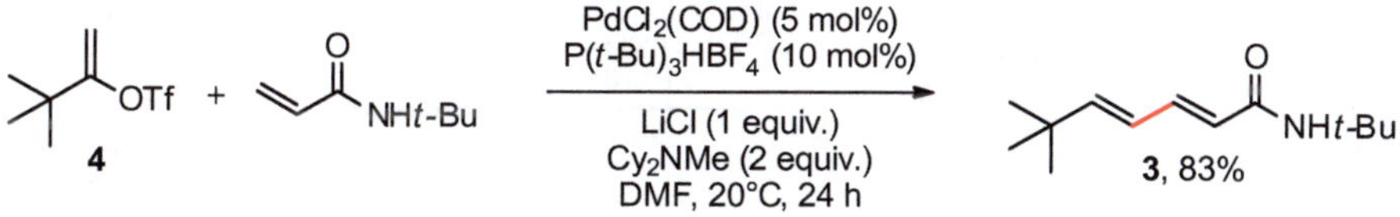

Scheme 2.3 The palladium 1,2-migration at room temperature. Reprinted (adapted) with permission from *J. Am. Chem. Soc.* **2007**, *129*, 6931. Copyright 2007 American Chemical Society

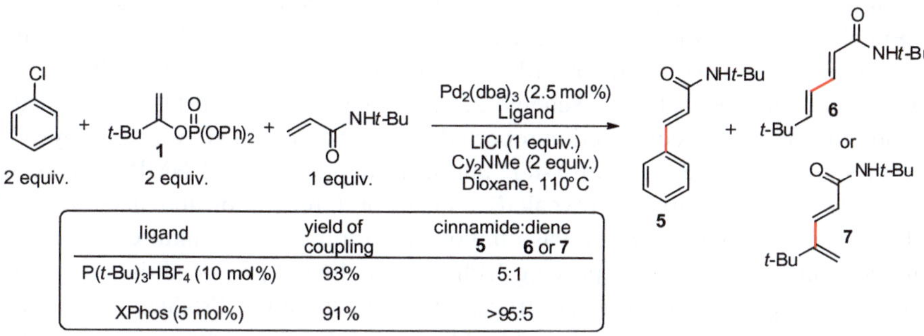

ligand	yield of coupling	cinnamide:diene 5 6 or 7
P(*t*-Bu)$_3$HBF$_4$ (10 mol%)	93%	5:1
XPhos (5 mol%)	91%	>95:5

Scheme 2.4 Competition experiment. Reprinted (adapted) with permission from *J. Am. Chem. Soc.* **2007**, *129*, 6931. Copyright 2007 American Chemical Society

Next, aryl vinyl phosphates carrying different substitution patterns were evaluated against *N-tert*-butylacrylamide using the developed migrating conditions (Table 2.4) [14]. In order to establish the influence of the electronic factors on the migratory process several aryl vinyl phosphates possessing both electron donating and withdrawing groups in the *para*-position were investigated. However, this resulted in complex crude reaction mixtures, and products arising from a possible Diels–Alder reaction between a migrated and a non-migrated diene were identified as the major product (results not shown). On the other hand, *ortho*-substituted aryl vinyl phosphates led to clean reactions, resulting in good to excellent yields although declined regioselectivities were obtained in some cases (entries 2–4). On the contrary, high levels of the migrated diene were obtained, when *o*-toluene or anthracene were introduced on the C1-position (entries 1 and 5). The electronic properties could explain the reluctance of the pentafluoroaryl vinyl phosphate to undergo 1,2-migration providing the non-isomerized Heck product exclusively (entry 6). Attempts to include the electron rich ferrocenyl substituent failed as the corresponding vinyl phosphate proved unstable (results not shown).

In order to test the influence of the temperature on the 1,2-migration process, a vinyl triflate was employed. The inherently high levels of reactivity, with respect to the oxidative addition, exhibited by triflate based electrophiles, typically allow the palladium catalyzed reactions to be conducted at lower temperatures. Hence, the Heck reaction of *tert*-butylvinyl triflate **4** and *N-tert*-butylacrylamide could successfully be accomplished at room temperature, applying the migrating reaction conditions, resulting in an 83 % isolated yield of **3**. More importantly, the

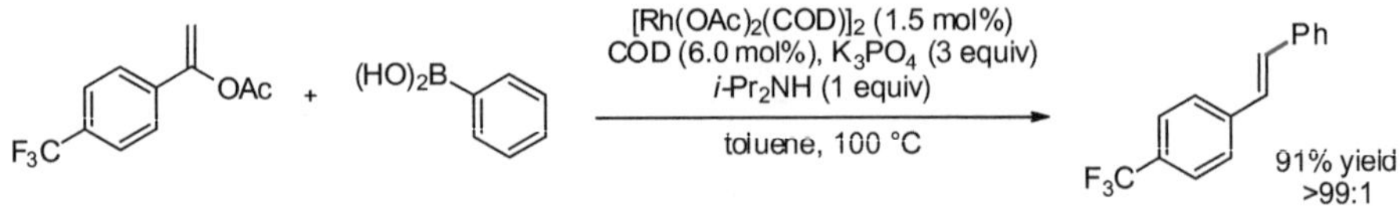

Scheme 2.5 Rhodium 1,2-migrations

migrated isomer was obtained exclusively, proving the alkenyl β-hydride elimination to be very efficient even at relatively low temperatures (Scheme 2.3).

Finally, a competition experiment between chlorobenzene and *tert*-butyl vinyl phosphate **1** was performed, in order to establish the level of reactivity of the involved alkenyl phosphates (Scheme 2.4). A catalytic system derived from the palladium catalyzed Heck reaction reported by Fu and coworkers was employed in the migratory Heck reaction of *N*-*tert*-butylacrylamide [15]. ^{1}H NMR analysis of the crude reaction mixture revealed a clear preference for the activation of chlorobenzene, when $P(t\text{-Bu})_3$ was employed, forming a 5:1 mixture of the cinnamide **5** and diene **6** or **7**. Applying XPhos as the ligand produced the cinnamide **5** as the sole product without any traces of C–O activation. Products originating from the oxidative addition of the aryl—O bond on the phosphate moiety were not detected during the Heck reaction of alkenyl phosphate demonstrating alkenyl phosphates to be more reactive than the aryl analogous. These findings situate the phosphate derived aryl and vinyl electrophiles at the end of the scale of reactivity, being less reactive than the aryl chlorides, with respect to the oxidative addition: $I > OTf > Br > Cl > OP(O)(OPh)_2$. Such information allows the chemist to design a specific catalyst capable of discriminating the sites of activation in systems carrying multiple reactive groups.

2.4 Conclusion and Outlook

Scope and limitation of the palladium 1,2-migration in the Heck reaction of alkenyl phosphates and electron deficient olefins yielding the isomerized dienes were further explored. Based on previous work performed within the group, a catalytic system composed of $PdCl_2(COD)$ and the sterically encumbered ligand $P(t\text{-Bu})_3$, employed as the air and moisture stable HBF_4 salt, in combination with stoichiometric amounts of LiCl and the tertiary amine base Cy_2NMe facilitated the 1,2-migration, exploiting different alkenyl phosphates. The steric properties and the nature of the C1 substituent proved crucial, revealing bulky alkyl and aryl groups most efficient in the formation of isomerized dienes. The underlying mechanism was proposed to involve a β-hydride elimination from a tricoordinated T-shaped alkenyl palladium(II) intermediate generating an alkyne coordinated palladium(II) species. Rotation of the alkyne followed by a carbopalladation, resulting in the 1,2-migrated alkenyl palladium(II) intermediate, would then ultimately lead to the isomerized dienes. In order to suppress the palladium

1,2-migration the electron rich XPhos ligand and 10 equivalents of LiCl were employed obtaining the non-isomerized Heck products.

Lately, Kuwano and coworkers reported on the 1,2-migration in the rhodium catalyzed reaction of alkenyl acetates and arylboron reagents, producing the isomerized *trans*-alkenes effectively. The catalytic system proved less sensitive to the steric interactions of the C1 substituent resulting in a broad scope and good regioselectivities (Scheme 2.5) [16].

References

1. Zhao J, Yue D, Campo MA, Larock RC (2007) J Am Chem Soc 129:5288
2. Mota AJ, Dedieu A (2006) Organometallics 25:3130
3. Li B–J, Yu D–G, Sun C–L, Shi Z–J (2011) Chem Eur J 17:1728
4. Hansen AL, Ebran J–P, Gøgsig TM, Skrydstrup T (2006) Chem Commun 4137
5. Hansen AL, Ebran J–P, Gøgsig TM, Skrydstrup T (2007) J Org Chem 72:6464
6. Gauthier D, Beckendorf S, Gøgsig TM, Lindhardt AT, Skrydstrup T (2009) J Org Chem 74:3536
7. Hansen AL, Skrydstrup T (2005) Org Lett 7:5585
8. Coe JW (2000) Org Lett 2:4205
9. Martin R, Buchwald SL (2008) Acc Chem Res 41:1461
10. Walker SD, Bareder TE, Martinelli JR, Buchwald SL (2004) Angew Chem Int Ed 43:1871
11. Barder TE (2006) J Am Chem Soc 128:898
12. Schnyder A, Indolese AF, Studer M, Blaser H-U (2002) Angew Chem Int Ed 41:3668
13. Amatore C, Jutand A (1999) J Organomet Chem 576:254
14. Hansen AL, Ebran J-P, Ahlquist M, Norrby P-O, Skrydstrup T (2006) Angew Chem Int Ed 45:3349
15. Littke AF, Fu GC (2001) J Am Chem Soc 123:6989
16. Yu J–Y, Shimizu R, Kuwano R (2010) Angew Chem Int Ed 49:6396

[illegible] complexation the electron rich XPhos ligand and 10 equivalents of [illegible] were [illegible] obtaining the [illegible] Heck [illegible] products.

[illegible]

References

[illegible]

Chapter 3
Palladium 1,2-Migration in the Negishi and Kumada Coupling

3.1 Introduction

In order to investigate and exploit the full potential of the palladium 1,2-migration, an implementation of the migratory process into the other named palladium catalyzed reactions would be of high interest. Consequently, complex products originating from simple starting materials could most certainly expand the scope and flexibility of palladium catalysis. For example, stilbenes, representing an important class of biological and pharmaceutical relevant compounds, are usually synthesized by a palladium catalyzed Heck reaction of aryl halides and styrenes [1]. Alternatively, these compounds could potentially be obtained by a palladium 1,2-migration process employing aryl vinyl phosphates in a Negishi or Suzuki coupling.

3.2 Background

During the studies on the implementation of alkenyl phosphates in the palladium catalyzed Negishi reaction producing 1,1-disubtituted alkenes, a new effective palladium 1,2-migration was discovered (Scheme 3.1)[2].

Preliminary results obtained from the coupling of *tert*-butylvinyl phosphate **8** and *o*-tolylzinc chloride **9** demonstrated high regioselectivity generating the isomerized *trans*-alkene exclusively, when the bidentate Josiphos ligand was employed. Additionally, the alkenyl tosylate **10**, possessing similar sterical properties, also facilitated the migrated product in good yields using the same reaction conditions. On the other hand, the non-migrated product was observed as the only species, when a catalytic system relying on $Pd_2(dba)_3$ and DPPF was exploited. Furthermore, $P(t\text{-}Bu)_3$, inducing the 1,2-migration effectively in the Heck reaction, proved inferior compared to Josiphos resulting in mixtures of the two isomeric alkenes.

T. M. Gøgsig, *New Discoveries on the β-Hydride Elimination*, Springer Theses,
DOI: 10.1007/978-3-642-32099-6_3, © Springer-Verlag Berlin Heidelberg 2012

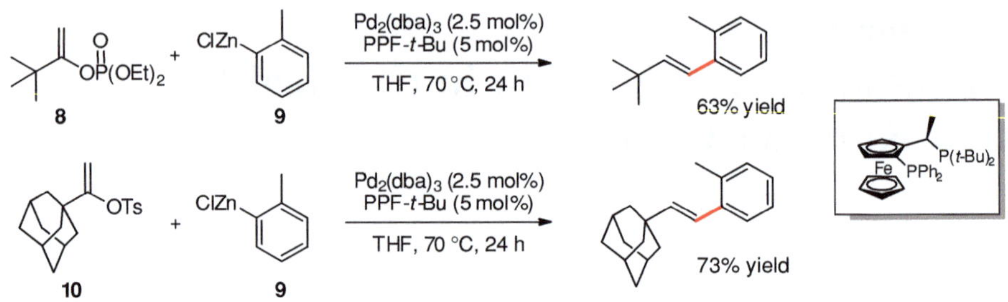

Scheme 3.1 Palladium 1,2-migration in the Negishi coupling

Based on DFT calculations and deuterium labeling experiments it was suggested, that the 1,2-migration proceeded through an alkenyl β-hydride elimination. However, in order to accomplish this unusual elimination process a free coordination site on the metal center is imperative and inherently attained using bulky trialkyl phosphines. On the contrary, four-coordinated square planar palladium(II) complexes are typically formed, when bidentate ligands are used, leaving a saturated coordination sphere. In particular, the Josiphos ligands have proved themselves extraordinary capable of stabilizing tetracoordinated palladium complexes attributable to the rigid ferrocene back bone, sterical properties and strong electron donating abilities [3].

Intrigued by the above presented results an in-depth investigation of the palladium 1,2-migration in the Negishi coupling was launched.

3.3 Results and Discussion

Initial investigations were performed using the alkenyl *O*,*O*-diphenyl phosphates adaptable in the Heck reaction. However, substantial amounts of unwanted byproducts such as homocoupling of the alkenyl phosphate were obtained, when these electrophiles were employed, and cleaner reactions were observed, when the corresponding *O*,*O*-diethyl phosphates and tosylates were introduced.

In order to examine the electronic and steric influence of the ligand on the 1,2-migration various commercially available Josiphos ligands holding phosphines with different characteristics were evaluated in the Negishi coupling between cyclohexylvinyl phosphate **11** and *o*-tolylzinc chloride **9** (Table 3.1). Although incomplete reactions were obtained in all cases, the results presented in Table 3.1 clearly demonstrate a significant ligand effect on the migratory process. Essentially full regioselectivity was observed, when NPF-*t*-Bu was tested possessing bulky *tert*-butyl groups on the methylene phosphine and naphthyl groups on the ferrocenyl phosphine (entry 5). As mentioned previously, high selectivity favoring the isomerized alkene product was achieved, when PPF-*t*-Bu was employed (entry 6). Introducing electron rich or deficient groups on the ferrocenyl phosphine did not

Table 3.1 Effects of the Josiphos ligands on the 1,2-migration

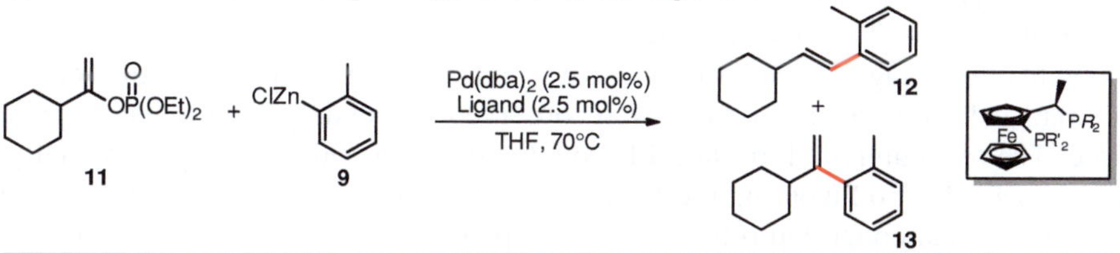

Entry	R′	R	Conversion[a] (%)	Ratio[a] 12:13
1	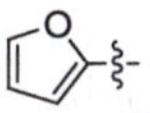	*t*-Bu	42	4:1
2	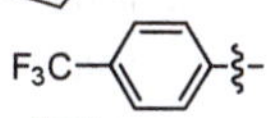	*t*-Bu	52	5:1
3	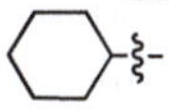	*t*-Bu	19	5:1
4	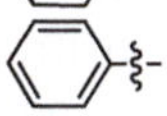	*t*-Bu	57	10:1
5	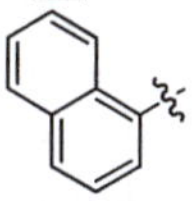	*t*-Bu	40	20:1
6	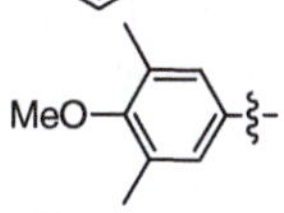	*t*-Bu	48	5:1
7	*t*-Bu	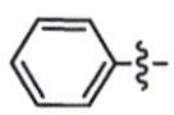	55	2:3
8	*t*-Bu	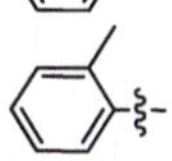	23	1:1

[a] Conversion and ratio determined by ^{1}H NMR analysis of crude reaction mixture. Reprinted (adapted) with permission from *J. Org. Chem.* **2009**, *74*, 135. Copyright 2009 American Chemical Society

provide any improvements (entries 1–3 and 6). Interestingly, 1:1 mixtures of the coupling products were obtained, when the two phosphines were exchanged suggesting distinct roles of the individual phosphines during the 1,2-migration (entries 7 and 8).

In general, alkenyl tosylates displayed enhanced reactivity compared to the corresponding phosphates in the Negishi cross couplings. However, decomposition of the aryl vinyl tosylates required the employment of the more stable aryl vinyl phosphates. Due to the weak coordinating abilities of the phosphate an alkenylpalladium(II) chloride complex is presumably formed after the oxidative addition similarly to the complex generated from the corresponding tosylate electrophiles [4].

Next, attention was directed towards the nucleophilic nature of the organozinc reagent potentially affecting the 1,2-migration. Various alkenyl phosphates and tosylates were coupled to arylzinc chlorides possessing different electronic and sterical properties. All reactions were conducted using the optimized catalytic system relying on PPF-*t*-Bu in THF at 70 °C for 18–24 h (Table 3.2). As illustrated in Table 3.2 the electron rich *para*-methoxy phenyl zinc chloride coupled smoothly resulting in good to excellent yields. However, modest ratios of the normal and isomerized products were observed, depending on the electrophile used. The *tert*-butylvinyl tosylate **14** provided the migrated product preferentially, whereas a 5:1 ratio favouring the 1,1-disubstituted alkene was obtained employing the less sterical congested *o*-tolyl vinyl phosphate **15** (entries 1–3). Replacing the electron donating methoxy group with the less donating methyl group led to somewhat lower yields, but more importantly enhanced the 1,2-migration slightly (entries 4 and 5). A more dramatic effect was detected, when changing from *p*-tolyl to *o*-tolyl. The hampered the nucleophilicity of the organometallic species resulted in good regioselectivities approaching full migration in the coupling of **14** (entry 6-8). On the other hand, the non-isomerized product was attained exclusively, when subjecting phenyl vinyl phosphate **17** to the palladium catalyzed negishi coupling (entry 9). Reducing the electron density in the aromatic ring by introducing a nitrile in the *para*-position improved the migratory process, although cyclohexylvinyl tosylate **16** demonstrated modest reactivity once again (entries 10–12). In order to test the influence of the coordinating properties of the leaving group on the 1,2-migration an arylvinyl chloride was subjected to the optimized reaction conditions utilizing *o*-tolylzinc chloride **9** as coupling partner (entry 13). The resulting *E*-stilbene derivative was obtained as the major product in a 5:1 ratio of the non-isomerized 1,1-disubstituted alkene in an overall 82 % isolated yield. A similar result was accomplished using the corresponding alkenyl phosphate suggesting identical key intermediates during the 1,2-migration process (entry 7).

Changing the ligand to the slightly more sterically hindered dinaphthyl derived Josiphos produced higher catalytic activity (Table 3.3). In general, good yields of the resulting stilbene derivatives were acquired, and better regioselectivities were observed, compared to the results obtained with diphenyl Josiphos. As depicted in Table 3.3 the migrated isomer was favoured in all cases except for the coupling of phenylvinyl phosphate (entry 9). However, a 1:2 mixture was generated in contrast to the 0:1 ratio using PPF-*t*-Bu (entry 9, Table 3.2).

Further investigations were performed increasing the nucleophilicity of the organometallic reagent and hence the rate of transmetallation. Accordingly, the regioselective outcome was studied employing alkylzinc and arylmagnesium reagents.

First, *n*-butylzinc chloride, readily obtained by mixing butyl lithium and zinc chloride, was evaluated in the palladium catalyzed Negishi coupling with 1,2-migration based on the catalytic system consisting of $Pd(dba)_2$ and PPF-*t*-Bu (Table 3.4). Subjecting the alkylzinc chloride to these conditions afforded the branched dialkyl olefin as the only product in excellent yields (entries 1 and 2). On the contrary, the more sterically hindered NPF-*t*-Bu ligated palladium complex afforded a 1:1 mixture of the two isomers in the coupling of *tert*-butylvinyl tosylate **14** albeit in lower yield (entry 3).

Next, the reaction temperature was investigated as illustrated in Table 3.5. From the results obtained in Table 3.5 it is evident, that higher levels of

Table 3.2 Palladium 1,2-migration in the Negishi couplings

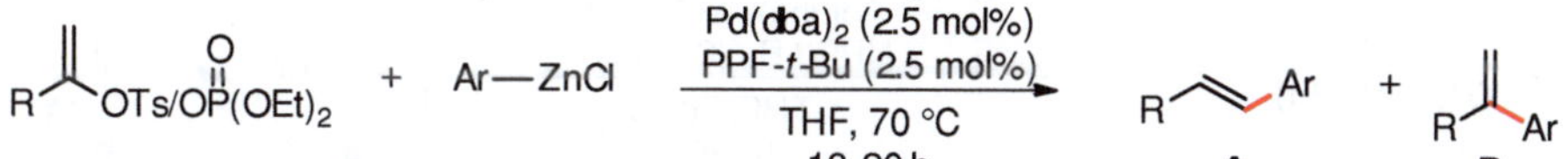

A B

Entry	Electrophile	Nucleophile	Conversion (%)[a]	Yield (%)[b]	Ratio A:B[a]
1	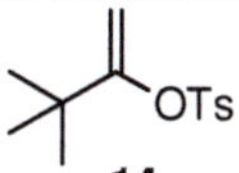 **14**	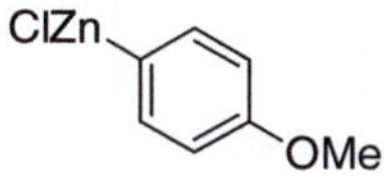 **19**	100	93	5:1
2	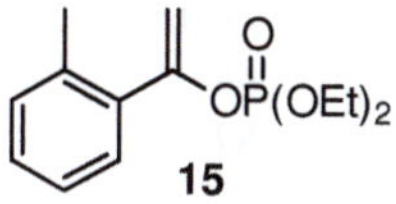 **15**	**19**	100	95	1:5
3	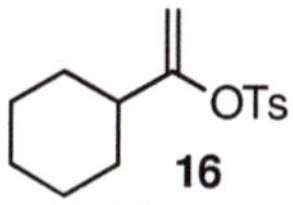 **16**	**19**	100	79	4:3
4	**14**	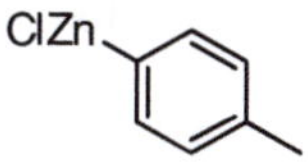 **20**	100	86	7:1
5	**16**	**20**	82	67	1:1
6	**14**	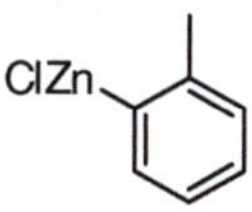 **9**	100	70	>19:1
7	**15**	**9**	95	72	9:2
8	**16**	**9**	39	n.d.[c]	5:1
9	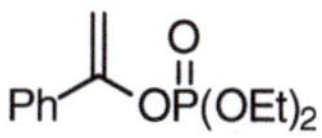 **17**	**9**	65	n.d.[c]	0:1
10	**14**	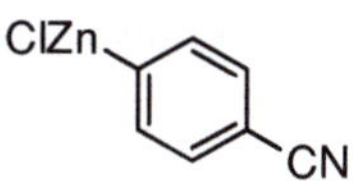 **21**	90	62	>19:1
11	**15**	**21**	60	59	5:1
12	**16**	**21**	32	n.d.[c]	10:1
13	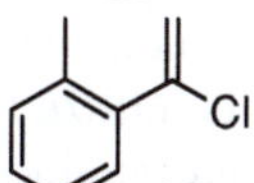 **18**	**9**	100	82	5:1

[a] Conversion and ratio determined by ^{1}H NMR spectrum of crude reaction mixture. [b] Isolated yields. [c] Not determined. Reprinted (adapted) with permission from *J. Org. Chem.* **2009**, *74*, 135.

Table 3.3 Palladium 1,2-migration in the Negishi couplings

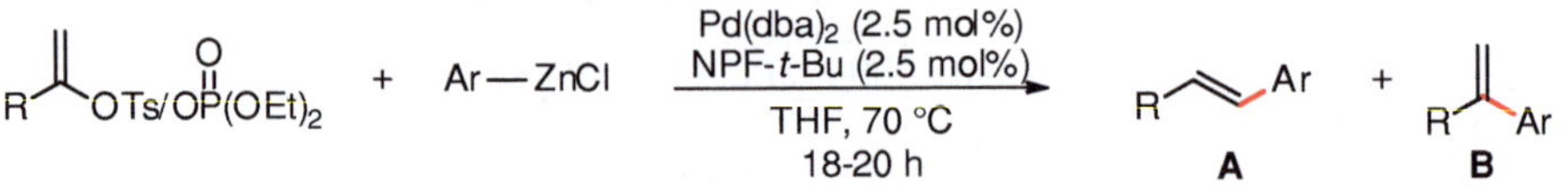

Entry	Electrophile	Nucleophile	Conversion (%)[a]	Yield (%)[b]	Ratio A:B[a]
1	**14** R = *t*-Bu	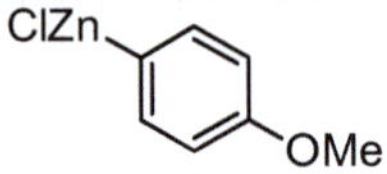 **19**	100	88	10:1
2	**15** R = *o*-tolyl	**19**	100	n.d.[c]	4:3
3	**16** R = Cy	**19**	100	67	2:1
4	**14** R = *t*-Bu	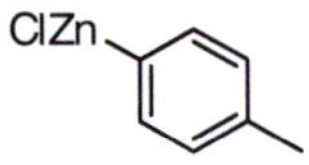 **20**	100	87	10:1
5	**16** R = Cy	**20**	100	71	5:2
6	**14** R = *t*-Bu	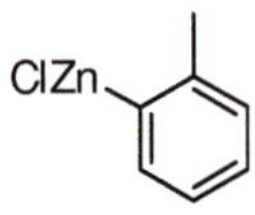 **9**	100	76	10:1
7	**15** R = *o*-tolyl	**9**	100	87	15:1
8	**16** R = Cy	**9**	92	70	5:1
9	**17** R = Ph	**9**	92	83	1:2
10	**14** R = *t*-Bu	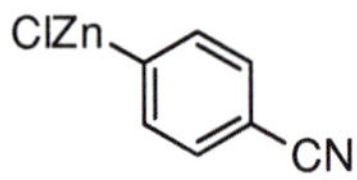 **21**	100	47	10:1
11	**15** R = *o*-tolyl	**21**	73	68	>19:1
12	**16** R = Cy	**21**	94	55	5:1

[a] Conversion and ratio determined by ^{1}H NMR spectrum of crude reaction mixture. [b] Isolated yields. [c] Not determined. Reprinted (adapted) with permission from *J. Org. Chem.* **2009**, *74*, 135. Copyright 2009 American Chemical Society

1,2-migration are generated by increased temperature demonstrating a significant temperature effect.

Finally, the difference between zinc and magnesium derived nucleophiles was studied introducing the highly nucleophilic Grignard reagents utilized in the Kumada reaction (Table 3.6). The non-isomerized products were attained in all cases, when a palladium catalyst derived from PPF-*t*-Bu was used. Nevertheless, short reaction times could be applied resulting in good to excellent yields (entries 1–4). As reported by Hartwig and coworkers, toluene proved to be the solvent of choice in the Kumada coupling of aryl and alkenyl tosylates and did not affect the 1,2-migration in the Negishi coupling as depicted in Scheme 3.2. On the other

Table 3.4 Palladium 1,2-migration in the Negishi couplings

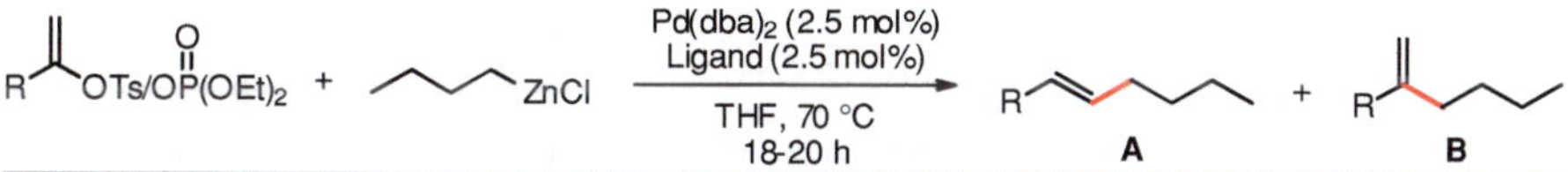

Entry	Electrophile	Ligand	Conversion (%)[a]	Yield (%)[b]	Ratio A:B[a]
1	**14** R = *t*-Bu	PPF-*t*-Bu	100	85	0:1
2	**15** R = *o*-tolyl	PPF-*t*-Bu	100	96	0:1
3	**14** R = *t*-Bu	NPF-*t*-Bu	100	58	9:10
4	**15** R = *o*-tolyl	NPF-*t*-Bu	100	86	0:1

[a] Conversion and ratio determined by ^{1}H NMR spectrum of crude product. [b] Isolated yields.-

Table 3.5 Temperature effect on the 1,2-migration

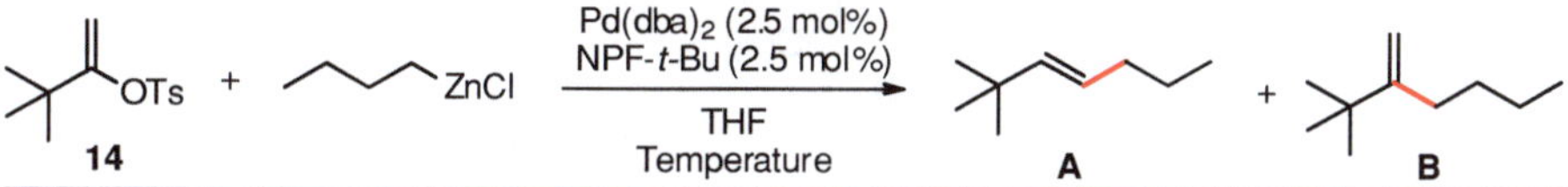

Entry	Temperature (°C)	Yield (%)[a]	Ratio A:B[b]
1	40	62	23:77
2	55	29	36:64
3	70	58	47:53

[a] Isolated yields. Ratio determined by ^{1}H NMR spectrum of crude product. [b] Isolated yields.

hand, THF lowered the catalytic turnover (results not shown).[5] Consequently the 1,2-migration in the Kumada coupling was investigated using toluene as solvent. The 1,2-migration observed by Hartwig and coworkers was achieved using the more electron donating and sterical congested ligand CyPF-*t*-Bu as depicted in Scheme 1.38. When a catalytic system relying on this ligand was exploited in the Kumada coupling of alkenyl tosylates, mixtures of the two isomers were isolated favouring the product originating from the 1,2-migration (entries 5 and 6). However, the normal non-isomerized product was once again formed as the only species, when the highly nucleophilic alkylmagnesium bromide was coupled to the alkenyl tosylate **7** (entry 7). Attempts to include the corresponding phosphates turned out unproductive as byproducts, formed by an apparent homocoupling of the alkenyl phosphates, were obtained [6].

3.4 Mechanistic Considerations

Based on the results presented in Tables 3.1, 3.2, 3.3, 3.4, 3.5 and 3.6 several distinct features can be identified as crucial in accommodating the palladium 1,2-

Table 3.6 Kumada couplings of alkenyl tosylates

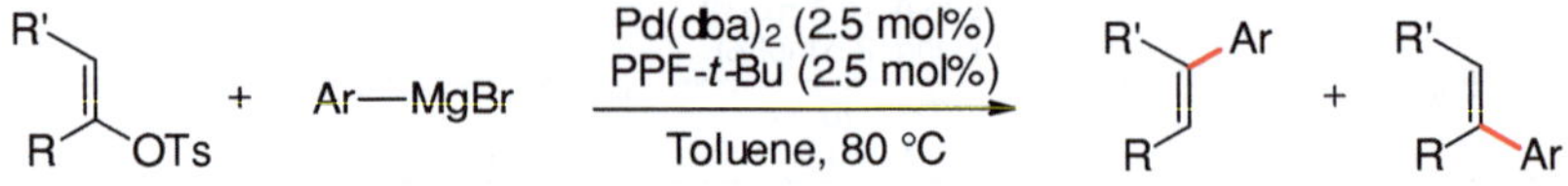

A B

Entry	Electrophile	Nucleophile	Time	Yield (%)[a]	Ratio A:B[b]
1	**14** R = *t*-Bu R′ = H	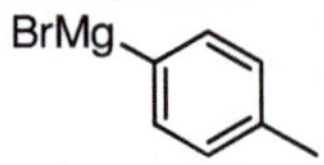 **23**	30 min	63	1:10
2	**16** R = Cy R' = H	**23**	35 min	91	0:1
3	**14** R = *t*-Bu R' = H	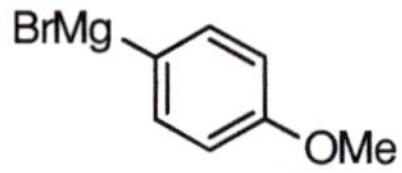 **24**	70 min	80	1:19
4	**22** R = *i*-Pr R' = Me	**24**	1.5 h	77	1:4
5	**14** R = *t*-Bu R' = H	**23**	30 min	59[c]	2:1
6	**14** R = *t*-Bu R' = H	**24**	70 min	32[c]	2:1
7	**22** R = *i*-Pr R' = Me	 **25**	45 min	87[c]	0:1

[a] Isolated yields. All reactions proceeded with full conversion. [b] Ratio determined by ^{1}H NMR spectrum of crude product. [c] The ligand CyPF-*t*-Bu (2.5 mol%) was employed. Reprinted (adapted) with permission from *J. Org. Chem.* **2009**, *74*, 135. Copyright 2009 American Chemical Society

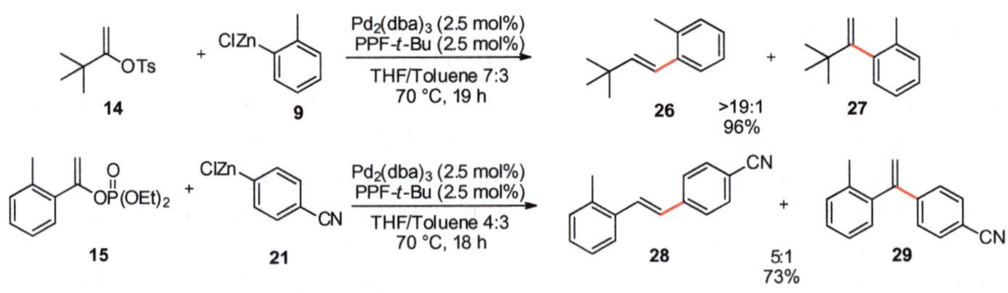

Scheme 3.2 Palladium 1,2-migration in the Negishi coupling. Reprinted (adapted) with permission from *J. Org. Chem.* **2009**, *74*, 135. Copyright 2009 American Chemical Society

migration. In the following a survey of the factors influencing the migratory process is presented.

From the results it can be concluded, that the nucleophilic character of the organometallic coupling partner has a significant influence on the 1,2-migration. The more nucleophilic coupling partners tend to suppress the migratory process, whereas the less reactive nucleophiles allow the isomerization to proceed.

In the Heck reaction it was established, that the regioselective outcome was profoundly dependant on the steric properties of the alkenyl tosylates and phosphates employed. Accordingly, a quaternary C1 substituent turned out mandatory in order to facilitate the vital alkenyl β-hydride elimination. Similar effects were observed in the Negishi and Kumada coupling, although a few examples of alkenyl electrophiles holding secondary alkyl groups proved suitable in the 1,2-migration as well.

Obviously the ligands used in the palladium 1,2-migrations hold a central position displaying a major influence on the regioselectivity. In the Heck reaction a highly electron donating and sterically encumbered trialkylphosphine was utilized generating a tricoordinated T-shaped palladium(II) intermediate after the oxidative addition holding a vacant site necessary for the β-hydride elimination to occur. In contrast, the 1,2-migration observed in the Negishi and Kumada couplings were attained using bidentate ligands. Bidentate ligands such as Josiphos and DPPF typically form saturated tetracoordinated palladium(II) complexes and hence leave the central metal unable to undergo alkenyl β-hydride elimination. This was indeed observed in the Negishi reaction of alkenyl phosphates using a catalytic system relying on DPPF [2]. However, it has been suggested, that DPPF may possess some monodentate character [7]. On the other hand, studies conducted by Hartwig and coworkers indicated, that the two phosphino groups present on the Josiphos ligands are strongly coordinated to the palladium nucleus thereby rendering monodentate properties of these ferrocene derived bisphosphines unlikely [3]. Nevertheless, a brief dissociation of one of the phosphines, allowing the β-hydride elimination to occur, cannot be excluded.

Alternatively, a cationic palladium(II) intermediate generated from the dissociation of the tosylate or phosphate subsequent to the oxidative addition could potentially be responsible for the observed 1,2-migration. Cationic palladium(II) complexes undergoing β-hydride eliminations are well described in the literature especially as key intermediates in the regioselective Heck reaction of electron rich olefins [8, 9]. In the presented Negishi reactions more than 40 equivalents of LiCl are present relative to the palladium catalyst and would most likely trap any cationic palladium(II) species generating a neutral palladium(II) chloride intermediate. This assumption was supported by the result obtained from the Negishi reaction of *o*-tolylvinyl chloride in Table 3.2 entry 13 providing the isomerized product predominantly.

The 1,2-migration could also be explained by a β-hydride elimination from the tetracoordinated alkenylpalladium(II) chloride complex generating a transient pentacoordinated palladium(II) intermediate, but the non-isomerized products obtained using DPPF as the auxiliary ligand rule this pathway out [2].

A plausible mechanism could possibly involve a brief dissociation of the least donating phosphino group generating a vacant site on the on the central

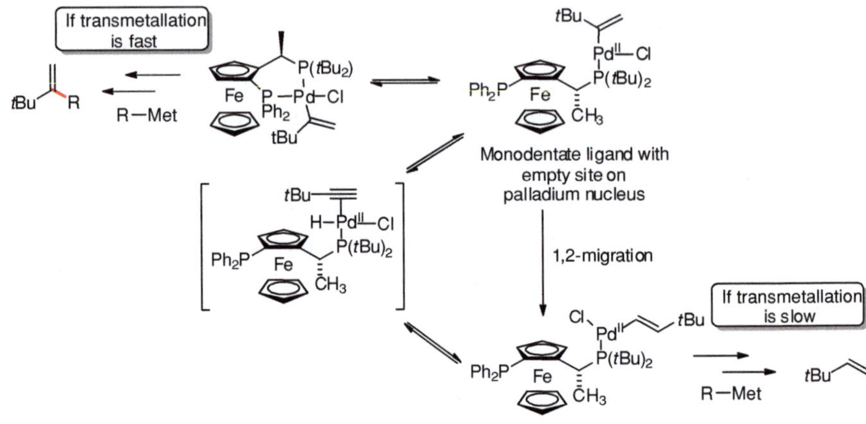

Scheme 3.3 Possible mechanism for the 1,2-migration in the Negishi coupling. Reprinted (adapted) with permission from *J. Org. Chem.* **2009**, *74*, 135. Copyright 2009 American Chemical Society

palladium(II) providing a trigonal palladium(II) complex similar to the complex suggested in the 1,2-migration in the Heck reaction (Scheme 3.3). The rate of transmetallation employing highly nucleophilic coupling partners could potentially outcompete the dissociation leading to the non-isomerized products. Thus, the rate of transmetallation is decisive with respect to the distribution of the migrated and non-migrated isomers.

The results obtained in the temperature experiments support this model. The transmetallation, representing a bimolecular process, would be expected to proceed to a greater extent at higher temperatures. This, however, was not observed implying that ligand dissociation becomes more feasible at elevated temperatures and hence favours the migratory process representing a unimolecular reaction.

3.5 Conclusion and Outlook

An effective palladium 1,2-migration was discovered in the Negishi and Kumada reactions, when a palladium catalyst based on the strongly electron donating and sterically congested Josiphos ligands were utilized. The rearrangement proved most effective for alkenyl tosylates and phosphates holding bulky substituents on the C1 position, similar to the observed results produced from the 1,2-migration in the Heck reaction. Furthermore, organometallic species displaying relatively low levels of nucleophilicity generated the isomerized products to a greater extent than highly reactive nucleophiles. Based on the obtained results a possible mechanism, involving the dissociation of the least donating phosphino group on the ligand leaving an open coordination site necessary for the crucial β-hydride elimination to occur, was proposed.

However, further investigations are needed in order to clarify the exact mechanism, as alternative pathways involving cationic intermediates or non-unimolecular processes cannot be expelled.

In the following stoichiometric studies on the palladium 1,2-migration conducted in the laboratories of Professor John F. Hartwig, University of Illinois, Urbana-Champaign are presented.

References

1. Ferré-Filmon K, Delaude L, Demonceau A, Noels AF (2004) Coord Chem Rev 248:2323
2. Hansen AL, Ebran J–P, Gøgsig TM, Skrydstrup T (2007) J Org Chem 72:6464
3. Shen Q, Ogata T, Hartwig JF (2008) J Am Chem Soc 130:6586
4. Hansen AL, Ebran J-P, Ahlquist M, Norrby P-O, Skrydstrup T (2006) Angew Chem Int Ed 45:3349
5. Limmert ME, Roy AH, Hartwig JF (2005) J Org Chem 70:9364
6. Gauthier D, Beckendorf S, Gøgsig TM, Lindhardt AT, Skrydstrup T (2009) J Org Chem 74:3536
7. Cabri W, Candiani I, DeBernardinis S, Francalanci F, Penco S, Santo R (1991) J Org Chem 56:5796
8. Ruan J, Iggo JA, Berry NG, Xiao J (2010) J Am Chem Soc 132:16689
9. Hansen AL, Skrydstrup T (2005) Org Lett 7:5585

However, further investigations are needed in order to clarify the exact mechanism as alternative pathways involving cationic intermediates or non-radical processes cannot be excluded.

In the following stoichiometric studies on the palladium(II) [illegible] [illegible]

References

[illegible]

Chapter 4
Stoichiometric Studies on the Palladium 1,2-Migration

4.1 Introduction

The β-hydride elimination from alkylpalladium(II) intermediates have been extensively described in the literature and exploited in numerous reactions as mentioned in Sect. 1.8. On the contrary, the analogous β-hydride elimination from alkenylpalladium(II) is essentially unprecedented, and studies dedicated to this elementary reaction are basically nonexistent.

4.2 Background

The oxidative addition of aryl tosylates to palladium(0) complexes carrying Josiphos ligands have been reported by Hartwig and coworkers, demonstrating this reaction to proceed through a neutral bend bisphosphinepalladium(0) complex (Scheme 4.1) [1, 2]. Similar to the studies reported by Amatore and Jutand the rate of oxidative addition was accelerated by added anions [3]. However, based on kinetic experiments the formation of an anionic palladium(0) species prior to the oxidative addition was excluded. Instead, the faster rates observed were attributed to an increase in the polarity of the medium. In particular, when low coordinating anions such as PF_6^- were added to the reaction mixture, increased rates of the oxidative addition were observed compared to the addition of strongly coordinating Br^- anions. Thus, the accelerating effects of added anions were ascribed a medium effect rather than the formation of an anionic palladium(0) species prior to the oxidative addition.

Treatment of the Pd(PPF-*t*-Bu)[P(*o*-tol)$_3$] **30** with phenyl tosylate in benzene at room temperature formed the resulting phenylpalladium(II) tosylate complex **32** in good yields via an initial dissociation of the P(*o*-tol)$_3$ ligand creating the more reactive bisphoshine palladium(0) species. The palladium(II) complex, holding the phenyl group cis to the less sterical hindered phosphino group, was produced as the

T. M. Gøgsig, *New Discoveries on the β-Hydride Elimination*, Springer Theses,
DOI: 10.1007/978-3-642-32099-6_4, © Springer-Verlag Berlin Heidelberg 2012

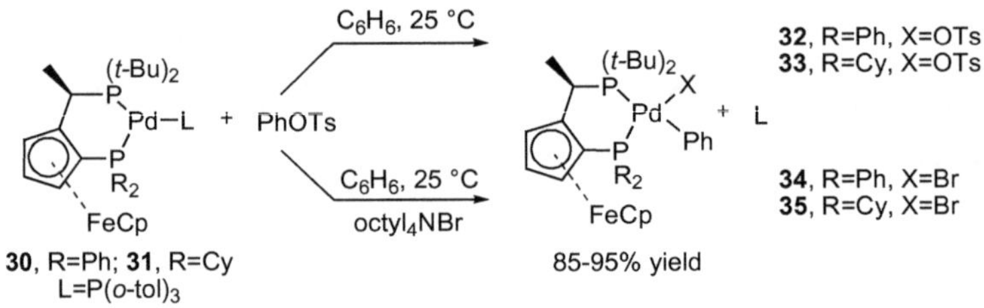

Scheme 4.1 Oxidative addition of aryl tosylates to Pd/Josiphos complexes. Reprinted (adapted) with permission from *J. Am. Chem. Soc.* **2003**, *125*, 8704. Copyright 2003 American Chemical Society

only isomer, which was verified by X-ray crystallography. The presence of bromide anions created the bromide substituted phenylpalladium(II) complex demonstrating the lower coordinating abilities of the tosylate.

Assuming that the alkenyl β-hydride elimination responsible for the 1,2-migration occurs after the oxidative addition and prior to the transmetallation, a similar approach, as the study presented in Scheme 4.1, could be applied in order to reveal the underlying mechanism of the palladium 1,2-migration in the Negishi and Kumada reactions. Hence, if the palladium 1,2-migration occurs by a unimolecular process, this would potentially be directly observable from the study of the oxidative addition of alkenyl electrophiles, possessing a bulky C1 substituent, to palladium(0) complexes carrying Josiphos ligands.

4.3 Results and Discussion

Initial investigations were conducted using the *tert*-butyl vinyl bromide **36** electrophile, since the resulting vinylpalladium(II) bromide complex were expected to be sufficiently stable and suitable for isolation and characterization. All manipulations were performed under an inert atmosphere using glovebox techniques. As demonstrated by Table 3.2, (p. 49) the leaving group did not influence the migratory process in the Negishi coupling, and comparable regioselectivities were obtained using the alkenyl chloride (entry 13). Hence, the vinyl bromide **36** employed during the preliminary studies was expected to undergo isomerization when subjected to the relevant catalyst, assuming the proposed mechanism depicted in Scheme 3.3 (p. 54) was operating.

The palladium(0) precursor Pd[P(o-tol)$_3$]$_2$ was obtained from Pd(dba)$_2$ and P(o-tol)$_3$ using known procedures [4]. Formation of Pd(PPF-t-Bu)[P(o-tol)$_3$] was accomplished by the stoichiometric reaction of Pd[P(o-tol)$_3$]$_2$ and PPF-t-Bu in THF and followed by ^{31}P NMR applying P(Mes)$_3$ as internal standard. Treating the resulting reaction mixture with the *tert*-butyl vinyl bromide **36** resulted in full conversion after only 10 min at room temperature, and a 72 % isolated yield of the non-migrated alkenyl Pd(PPF-t-Bu)(Br) could be secured (Scheme 4.2).

Scheme 4.2 Non-migrated alkenylpalladium(II) bromide complex **37**

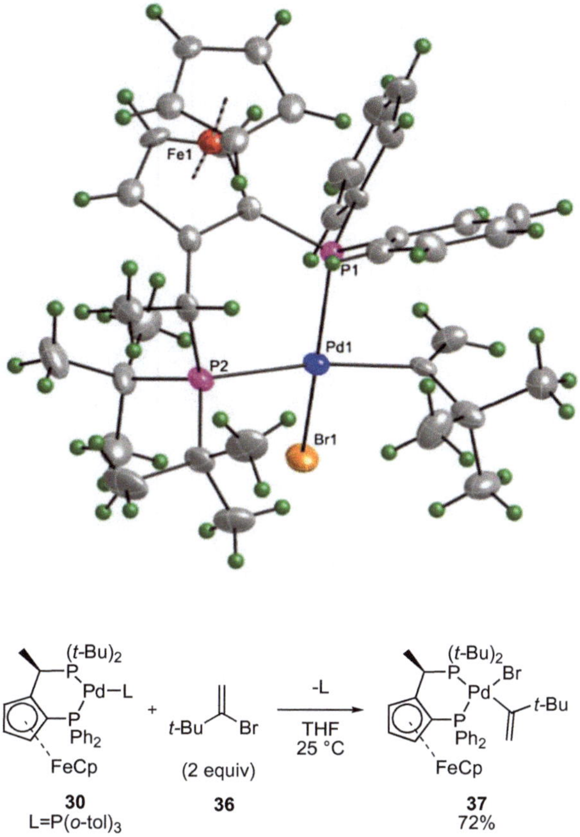

Surprisingly, the migrated species was not detected. In order to investigate the pronounced temperature effects, illustrated by the results presented in Table 3.5 (p. 51), temperature experiments were performed testing different solvents (Table 4.1). The existence of a *trans*-double bond was identified by $^{1}H\{^{31}P\}$- and ^{1}H NMR of the crude reaction mixture. However, in order to determine the exact structure of the palladium migrated complex **38**, isolation and characterization of the vinylpalladium(II) bromide complex **38** is mandatory.

The results presented in Table 4.1 clearly demonstrate the expected influence of the temperature on the migratory process. Higher levels of 1,2-migration were observed in all solvents, when elevated temperatures were applied to the reaction mixture. Interestingly, the Negishi reactions were performed in THF at 70 °C achieving good regioselectivities in favour of the migrated product, whereas 110 °C was needed in order to promote the 1,2-migration in the stoichiometric experiments (entry 3). In addition, the polarity of the solvent revealed itself of utmost importance. Full migration was attained in DMF at 80 °C after only 10 min, whereas a 2:1 mixture favouring the non-migrated alkenylpalladium(II) complex was obtained in toluene after 290 min(entries 5 and 8). These findings

Table 4.1 Temperature effect on the 1,2-migration

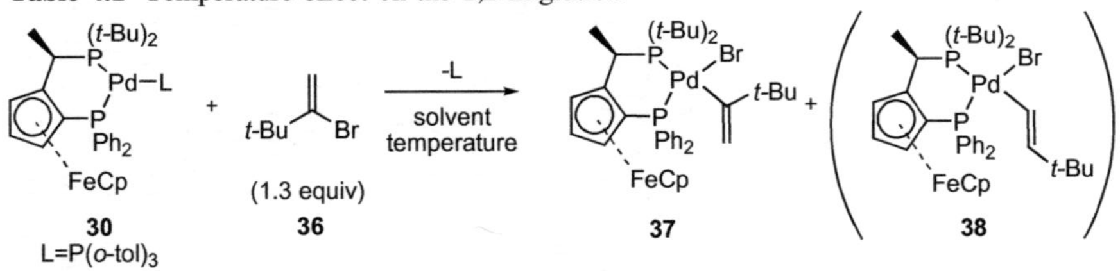

Entry	Solvent	Temperature (°C)	Time (min.)	Yield (%)[a]	Ratio **37**:**38**[b]
1	THF-d_8	45	135	55	1:0
2	THF-d_8	70	30	–[c]	1:0
3	THF-d_8	110	10	54	0:1
4	Toluene-d_8	45	290	55	1:0
5	Toluene-d_8	80	290	54	2:1
6	Toluene-d_8	110	90	43	0:1
7	DMF	45	210	88	2:5
8	DMF	80	10	45	0:1
9	DMF	110	10	45	0:1

[a] Yield determined by ^{31}P NMR spectroscopy using internal standard. [b] Ratio determined by ^{1}H NMR spectrum of crude product. [c] Decomposition

Scheme 4.3 Migrated alkenylpalladium(II) bromide complex **38**

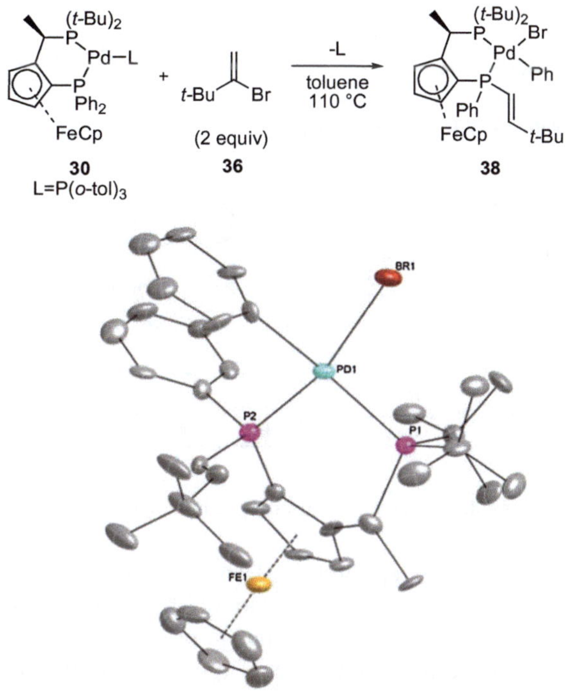

deviate from the observations made in the Negishi reaction conducted in THF:toluene mixtures achieving a highly effective 1,2-migration (Scheme 3.2, p. 52).

Based on the obtained results the migrated alkenylpalladium(II) complex was synthesized by adding *tert*-butyl vinyl bromide **36** to a toluene solution containing the highly reactive Pd(PPF-*t*-Bu)[P(*o*-tol)$_3$] and reacted at 110 °C for 2 h (Scheme 4.3). As illustrated in the ORTEP diagram, an interchange of the migrated alkenyl and one of the phosphino phenyl groups has occurred. Although the structure of the complex is different from the expected migrated alkenylpalladium(II) bromide complex, it still serves as the first strong evidence for a unimolecular palladium 1,2-migratory process most likely involving an alkenyl β-hydride elimination. Hence, the results presented in Table 4.1 are still valid, reflecting the influence of the temperature on the palladium 1,2-migration preceding the interchange of the phenyl and alkenyl groups.

The C—P bond cleavage processes responsible for the observed interchange has been described in the literature. Novak and coworkers reported on the reversible formation of tetraarylphosphonium ion, via a reductive elimination pathway, as a possible key intermediate in the exchange of aryl groups on the central palladium with aryl groups on the phosphine moiety (Scheme 4.4) [5]. A subsequent oxidative addition of a different C–P bond in the phosphonium ion to the intermediary palladium(0) complex would then create the interchanged arylpalladium(II) complex also observed by Chernard and coworkers [6]. A similar C–P bond

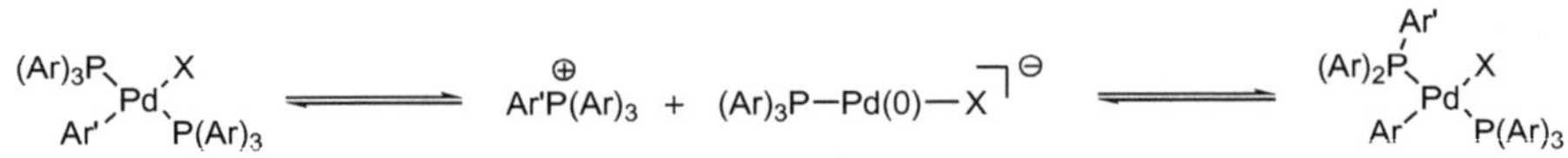

Scheme 4.4 C–P bond cleavage

Scheme 4.5 C–P bond cleavage using bidentate ligands

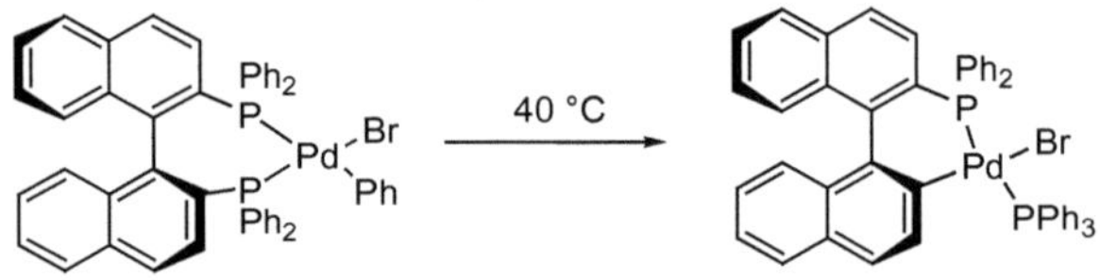

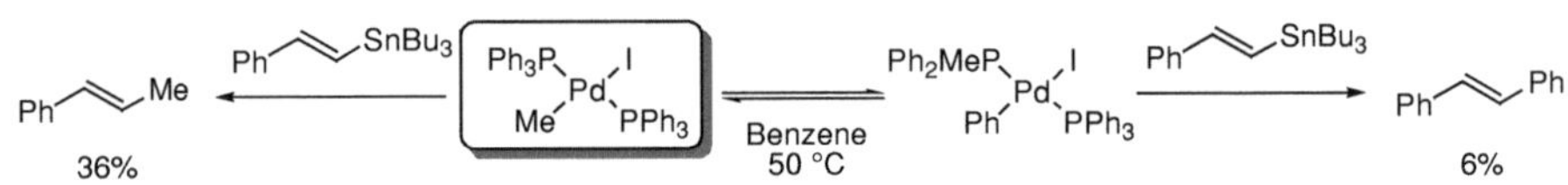

Scheme 4.6 Interchange of alkyl and phenyl groups

cleavage was reported by Hartwig and coworkers applying bidentate ligands such as BINAP and DPPF at relatively low temperatures (Scheme 4.5) [7].

Furthermore, the interchange of a methyl group on the central palladium and one of the phosphino phenyl groups on PPh_3 was detected in the stoichiometric Stille coupling of β-styryl stannane and a *trans*-phenylpalladium(II) iodide complex by the group of Norton (Scheme 4.6) [8]. Although the stilbene product was obtained in low yields, this example demonstrates the capability of alkyl groups to participate in the C–P bond cleavage processes.

In order to resemble the 1,2-migration process in the Negishi coupling to a greater extent oxidative addition of the *tert*-butyl vinyl tosylate **14** was investigated using the Pd/PPF-*t*-Bu system applied in the Negishi reactions. Suitable reaction conditions for the oxidative addition of alkenyl tosylate **14** to Pd(PPF-*t*-Bu)[P[*o*-tol)$_3$] were quickly identified being similar to the conditions reported by Hartwig and coworkers (Table 4.2) [1, 2]. In contrast to the *tert*-butyl vinyl bromide **36** the *trans*-alkene was identified by $^1H\{^{31}P\}$ NMR and ^1H NMR at room temperature after relatively short reaction times independent of the solvent used. Furthermore, high polar solvents were not required in order to promote the 1,2-migration.

In order to validate the structure of **40** isolation and characterization of the migrated palladium(II) complex were needed. To Pd(PPF-*t*-Bu)[P[*o*-tol)$_3$] dissolved in benzene was added *tert*-butylvinyl tosylate **14** and reacted at 35 °C. ^{31}P NMR showed complete reaction after 30 min and the resulting palladium(II) complex could be secured upon filtration through Celite followed by precipitation from a concentrated benzene solution using pentane. However, crystals suitable for X-ray crystallography were very difficult to obtain. Only after tedious experimentation it was found, that needles suitable for X-ray crystallography could be obtained by recrystallization from a concentrated toluene solution by diffusion of pentane at −20 °C for 7 days (Scheme 4.7). Nevertheless, a similar C–P bond

Table 4.2 Palladium 1,2-migration of alkenyl tosylate

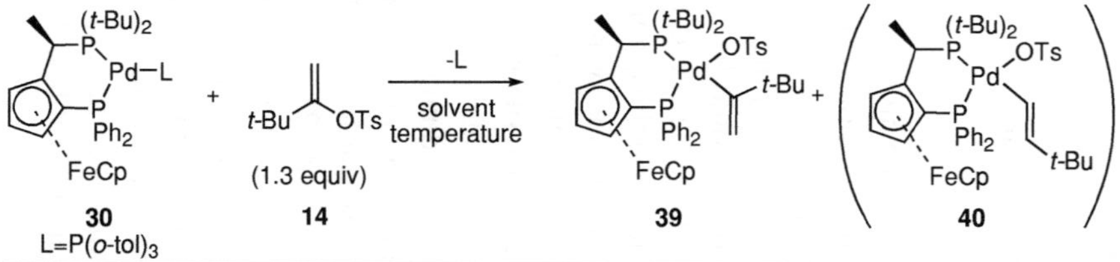

Entry	Solvent	Temperature (°C)	Time (min.)	Yield (%)[a]	Ratio **39**:**40**[b]
1	C_6D_6	25	90	71	0:1
2	THF-d_8	25	150	52	0:1
3	THF-d_8	45	30	50	0:1
4	THF-d_8	80	<5	48	0:1

[a] Yield determined by ^{31}P NMR spectroscopy using internal standard. [b] Ratio determined by ^{1}H NMR spectrum of crude product

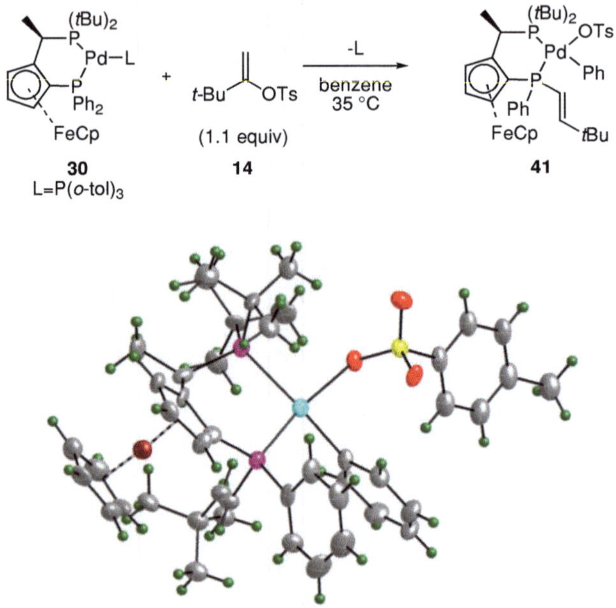

Scheme 4.7 Migrated alkenylpalladium(II) tosylate complex **41**

cleavage was once again observed yielding the phenylpalladium(II) complex as the only species. In both cases the migrated alkenyl moiety was positioned on the less hindered phosphine in agreement with the findings reported by Novak and coworkers, demonstrating bulky electron rich phosphines to hamper the C–P bond cleavage reaction [5]. Potentially, this process could have a detrimental effect in palladium catalyzed vinylation reactions changing the catalytic properties of the central palladium during the reaction, when arylphosphines are employed. However, in the presented Negishi reaction good yields were secured in a number of cases, which could possibly be explained by competing rates of the transmetallation compared to the C–P bond cleavage. Alternatively, the interchanged palladium species could perhaps act as the active catalyst.

Even though an interchanged phenylpalladium(II) tosylate complex was obtained, the 1,2-migration proved to be very fast even at room temperature complementing the 1,2-migration from tricoordinated T-shaped palladium(II) complexes observed in the Heck reaction (Chap. 2).

From the results presented in Tables 4.1 and 4.2 it is evident, that the leaving group has a pronounced effect on the migratory process contrasting the results obtained in the 1,2-migration in the Negishi reaction, which showed that alkenyl tosylates, phosphates and chlorides were equally capable of promoting the palladium 1,2-migration. Furthermore, the addition of halides to the reaction mixture turned out devastating for the 1,2-migration (Table 4.3). In the presence of 5

Table 4.3 Halide effect on the palladium 1,2-migration of alkenyl tosylate

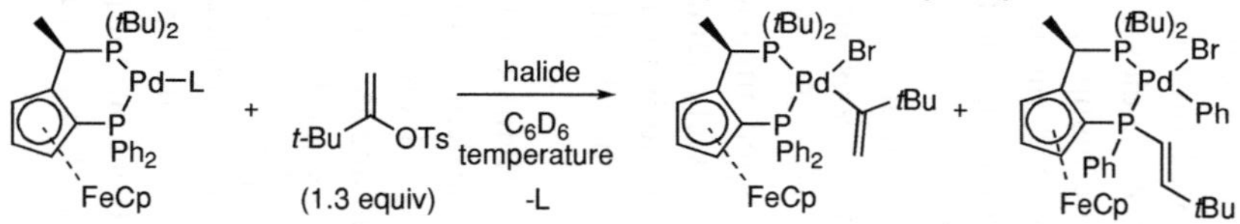

30 L=P(o-tol)$_3$ 14 37 38

Entry	Halide source	Temperature (°C)	Yield (%)[a]	Ratio **37**:**38**[b]
1	TOAB	25	68	1:0[c]
2	TOAB	80	57	5:8[c]
3	TOAB	25	27	1:0[d]
4	TBAC	25	77	1:0[e]
5	TBAC	50	56	1:0[e]
6	TBAC	80	36	1:1[e]

[a] Yield determined by ^{31}P NMR spectroscopy using internal standard after 24 h. [b] Ratio determined by ^{1}H NMR spectrum of crude product. [c] TOAB (tetraoctylammonium bromide) 5 equiv. [d] TOAB 1.5 equiv. [e] TBAC (tetrabutylammonium chloride) 5 equiv.

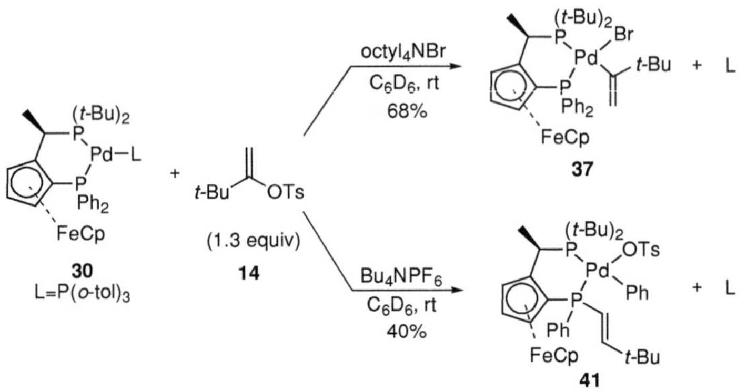

Scheme 4.8 Medium effect on the palladium 1,2-migration

equivalents of tetraoctylammonium bromide the non-migrated alkenylpalladium(II) bromide complex **37** was formed exclusively proving the poor coordinating abilities of the tosylate leaving group (entry 1). Elevating the temperature to 80 °C enhanced the 1,2-migration although to a modest extent (entry 2). Conducting the reaction with only 1.5 equivalents of TOAB resulted similarly in the non-migrated alkenylpalladium(II) bromide complex **37** (entry 3). However, a lower yield was obtained in this case indicating a stabilizing effect of the elevated ionic strength of the medium. The migratory process was even more hampered in the presence of chloride ions resulting in a 1:1 mixture of the migrated and non-migrated species compared to a 5:8 ratio favouring the migrated palladium intermediate after 24 h at 80 °C as determined by ^{1}H- and ^{1}H{^{31}P} NMR.

In order to establish whether the ionic strength medium was responsible for the inhibition of the migratory process an experiment was performed adding 5 equivalents of the weak coordinating PF_6^- (Scheme 4.8). Full conversion was observed after only 15 min forming the interchanged migrated alkenylpalladium(II) tosylate complex **41** exclusively in a 40 % yield relative to the internal standard excluding the ionic strength of the medium to inhibit the 1,2-migration. Instead, this result demonstrates the nature of the anion to be of utmost importance in the obstruction of the migratory process.

The proposed mechanism for the 1,2-migration relying on a dissociation of the least donating phosphino group generating a transient neutral trigonal palladium(II) complex is inconsistent with the observed solvent and ionic effects. The marked influence of the nature of the leaving group and the presence of halides on the migratory process cannot be justified using this model (Scheme 3.3, p. 54). Thus, from the results presented so far in this chapter, an intermediary cationic palladium(II) complex seems to be responsible for the palladium 1,2-migration employing Josiphos ligands. However, the presence of halide ions did not influence the palladium 1,2-migration in the Negishi reaction but proved to exhibit a detrimental effect in the stoichiometric investigations trapping the cationic

Scheme 4.9 Zinc chloride as halide scavenger

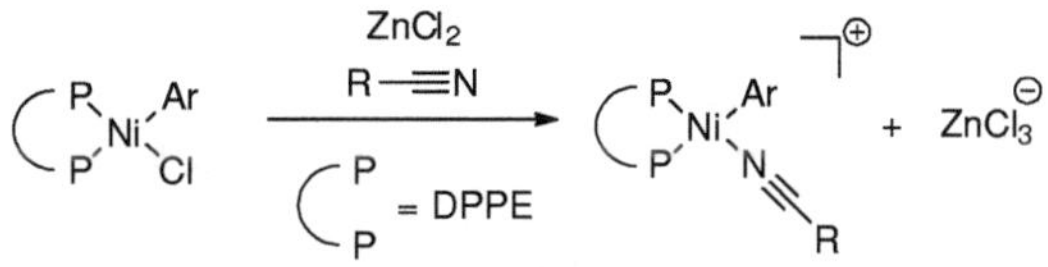

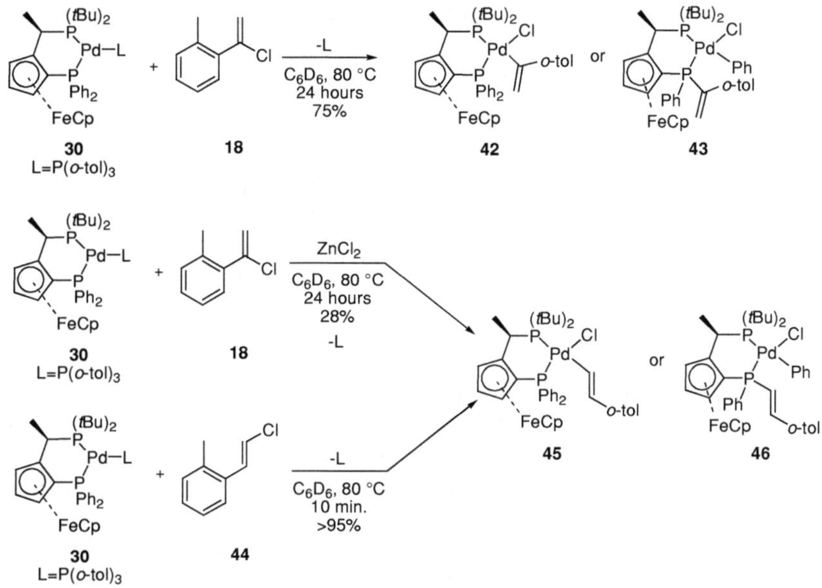

Scheme 4.10 Zinc mediated palladium 1,2-migration

palladium(II) complex. Hence further investigations were needed in order to clarify the exact mechanism of the palladium 1,2-migration in the Negishi reactions.

Dissociation of halides leaving a cationic palladium(II) complex has been reported by Xiao and others employing high polar solvents or halides scavengers such as silver or thallium salts [9–11]. In the Negishi reactions large amounts of divalent zinc are present, which could possibly act as halide scavenger generating the postulated cationic palladium(II) intermediate. Recently, Cheng and coworkers demonstrated divalent zinc ions to act as halide scavengers in the nickel catalyzed addition of arylboronic acids to nitriles (Scheme 4.9) [12].

Two experiments were conducted reacting the arylvinyl chloride **18** and the Pd(PPF-*t*Bu)[P(*o*-tol)$_3$] **30** in C_6D_6 at 80 °C (Scheme 4.10, see supporting information). Without any additives the non-migrated alkenylpalladium(II) chloride complex was obtained in a 75 % yield after 24 h as determined by ^{31}P NMR using P(Mes)$_3$ as internal standard. Changing to THF or DMF at elevated temperatures did not facilitate the 1,2-migration to any detectable degree (results not shown). On the other hand, adding 5 equivalents of zinc chloride led to a palladium(II) species holding a ^3J (H, H) *trans* double bond indicating, that a zinc mediated

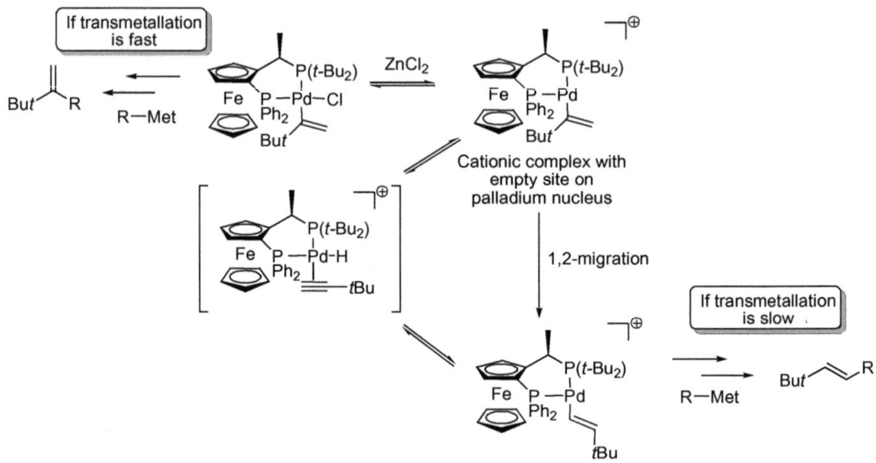

Scheme 4.11 Plausible mechanism of the palladium 1,2-migration in the Negishi reaction

palladium 1,2-migration had occurred. Furthermore, employing THF as the solvent produced the same result after only 10 min at 70 °C indicating an efficient 1,2-migration, when more polar solvents were used (results not shown). In order to confirm the observed ^{1}H and ^{31}P chemical shifts, the migrated palladium(II) species was independently synthesized starting from *E*-1-(2-chlorovinyl)-*o*-toluene **44**. Identical ^{1}H and ^{31}P chemical shifts were attained verifying the beneficial effects of zinc chloride on the palladium 1,2-migration. However, the exact structure of this complex was not clarified and an interchange of the vinyl group on the palladium center and the phenyl on the phosphino group cannot be ruled out.

Based on these results a modified mechanistic pathway for palladium 1,2-migration in the Negishi reaction can be rationalized (Scheme 4.11).

Due to the weak coordinating abilities of the tosylate a sterically congested tetracoordinated alkenylpalladium(II) chloride is generated after the oxidative addition. The large quantities of divalent zinc ions present in solution facilitate the formation of the cationic palladium(II) intermediate by a zinc mediated chloride abstraction. Subsequently, the unsaturated cationic alkenylpalladium(II) species undergoes a *β*-hydride elimination generating a cationic alkyne coordinated palladium(II) hydride complex, which upon rotation of the alkyne followed by a hydropalladation forms the isomerized and less sterically encumbered alkenylpalladium(II) complex. Provided that the rate of transmetallation is sufficiently low, the isomerized 1,2-disubstituted alkenes are produced as in the case of the Negishi couplings. If the rate of transmetallation exceeds the formation of the cationic palladium(II) complex the non-migrated 1,1-disubstituted alkenes are obtained as illustrated by the Kumada couplings.

In the efforts to avoid the C–P bond cleavage in the palladium 1,2-migration a number of ligands were investigated including DPPF, D*i*PrPF, P(*t*-Bu)$_3$ and P(*o*-tol)$_3$. The migrated alkenylpalladium(II) complex could not be identified using these ligands demonstrating the specific requirements needed in order to promote the palladium 1,2-migration. More promising results were obtained using CyPF-*t*-Bu as ligand enabling the 1,2-migration in the Kumada coupling (Table 3.6, p. 52). The presence of the dicyclohexyl moiety on the ferrocenyl phosphine would most likely prevent the C–P bond cleavage due to the more electron rich and hence stronger C–P bond. The palladium complex obtained with this ligand would therefore serve as a more stable alternative to the PPF-*t*-Bu derived alkenylpalladium(II) complex. However, initial attempts to isolate and characterize the involved intermediates, using a similar approach as reported by Hartwig and coworkers, turned out unsuccessful (Scheme 4.1). Further investigations and kinetic studies employing the CyPF-*t*-Bu ligand are still needed in order to uncover the exact mechanism of the palladium 1,2-migration in the Negishi and Kumada couplings.

4.4 Conclusion and Outlook

Based on the results obtained so far the palladium 1,2-migration most likely proceeds through an unprecedented alkenylpalladium(II) β-hydride elimination. A number of factors were identified to affect the 1,2-migration, all of which indicated the involvement of a central cationic alkenylpalladium(II) intermediate.

The nature of the leaving group was found to hugely influence the migratory process. Alkenyl tosylates facilitated the 1,2-migration more readily than the corresponding alkenyl bromides and chlorides generating the migrated palladium complex at room temperature in non-polar solvents such as benzene. On the other hand, high temperatures and prolonged reaction times were required in order to promote the 1,2-migration from alkenylpalladium(II) halide complexes. Alternatively, polar solvents such as THF or DMF could be applied producing the migrated species more efficiently. These observations could be explained by the reluctance of halides to dissociate from the palladium nucleus forming the cationic palladium(II) intermediate more slowly compared to the tosylate analogue displaying weaker coordinating abilities. In particular, this was illustrated by the obstruction of the migratory process, when halide anions were added to the reaction mixture of *tert*-butyl vinyl tosylate **14**. Notably, the presence of divalent zinc circumvented this situation acting as a halide scavenger accommodating the formation of the central trigonal cationic palladium(II) intermediate.

These findings are likely to account for the observed 1,2-migrations obtained in the palladium catalyzed Negishi couplings relying on the Josiphos ligands and constitute the first strong indications for an alkenyl β-hydride elimination generating the key alkyne coordinated palladium(II) intermediate, which ultimately leads to the 1,2-migrated products.

In the Heck 1,2-migration utilizing the sterically encumbered $P(t\text{-}Bu)_3$ ligand the situation is fundamentally different. In this case, a free coordination site on the central palladium is inherently present due to the T-shaped tricoordinated palladium(II) complex allowing the 1,2-migration to proceed without any preceding phosphine or counterion dissociation.

The full potential and applications of the palladium 1,2-migrations are still to be explored.

In the following transition metal catalyzed reactions relying on β-hydride eliminations from cationic organometallic intermediates are presented.

References

1. Roy AH, Hartwig JF (2003) J Am Chem Soc 125:8704
2. Roy AH, Hartwig JF (2004) Organometallics 23:194
3. Amatore C, Jutand A, Suarez A (1993) J Am Chem Soc 115:9531
4. Ogata T, Hartwig JF (2008) J Am Chem Soc 130:13848
5. Goodson FE, Wallow TI, Novak BM (1997) J Am Chem Soc 119:12441
6. Segelstein BE, Butler TW, Chenard BL (1995) J Org Chem 60:12
7. Alcazar-Roman LM, Hartwig JF, Rheingold AL, Liable-Sands LM, Guzei IA (2000) J Am Chem Soc 122:4618
8. Morita DK, Stille JK, Norton JR (1995) J Am Chem Soc 117:8576
9. Sonesson C, Larhed M, Nyqvist C, Hallberg A (1996) J Org Chem 61:4756
10. Ruan J, Xiao J (2011) Acc Chem Res 44:614
11. Arvela RK, Pasquini S, Larhed M (2007) J Org Chem 72:6390
12. Wong Y–C, Parthasarathy K, Cheng C–H (2010) Org Lett 12:1736
13. Nicolau KC, Sorensen EJ (2003) Classics in total synthesis; Wiley-VCH, Weinheim

Chapter 5
Heteroaromatic Tosylates in the Regioselective Heck Reaction

5.1 Introduction

Regioselective Heck reactions of electron rich olefins are as stated above traditionally conducted using aryl triflates as electrophiles. The formation of the crucial cationic palladium(II) complex is readily formed after the oxidative addition due to the weak coordinating abilities of the triflate anion. The employment of other sulfonate derivatives such as aryl tosylates in the Heck reaction, representing a more stable and less costly alternative to the aryl triflates, has hardly been described in the literature. Although these electrophiles have been utilized in numerous palladium catalyzed reactions including Suzuki, [1] Sonogashira, [2] Kumada, [3] α-arylations, [4] Buchwald-Hartwig aminations [5] etc., the different catalytic cycle and higher requirements to the catalytic system most likely account for their unexploited potential in the Heck reaction. The relatively low coordinating properties of the tosylate anion to the central palladium most likely facilitate the formation of the cationic palladium(II) intermediate, when a bidentate ligand in combination with an appropriate solvent are employed. Hence, aryl tosylates may serve as an attractive entry point to the regioselective Heck reaction of electron rich olefins [6].

In the following, the regioselective Heck reaction of heteroaromatic tosylates and electron rich olefins will be presented and discussed. This protocol constitutes the first general strategy for the palladium catalyzed olefination of pyridyl tosylates obtaining perfect α-regioselectivity.

5.2 Background

It has been estimated, that around one-third of all organic compounds constructed in the drug development programs throughout the pharmaceutical industry include heteroaromatic fragments [7]. The introduction of a heteroaromatic moiety

T. M. Gøgsig, *New Discoveries on the β-Hydride Elimination*, Springer Theses,
DOI: 10.1007/978-3-642-32099-6_5,

significantly changes the pharmacodynamic and pharmacokinetic properties of the drug compared to the corresponding benzene analogous. Such changes in the pharmacological effects may originate from molecular interactions through hydrogen bonding, metal coordination, modified solubility, altered polarization etc. [8]. As a result, a vast number of strategies focusing on the syntheses of privileged structures such as pyridine, [9] pyrimidine, [10] quinazoline, [11] pyrazole [12] and so forth have been reported in order to meet the demands for increased structural diversity. Furthermore, the functionalization of heteroaromatic systems has attracted much attention and has often been achieved by strategies relying on transition metal catalysis [13, 14, 15]. However, coordination of the heteroatom to the central metal may lead to catalyst inhibition, which complicates the design of a flexible and general catalytic system [16].

Recently, Skrydstrup and coworkers reported on the implementation of 2-hydroxypyridines and pyrimidines, relying on sulfonates different from triflates, in the transition metal catalyzed C–C and C–N bond forming reactions obtaining the functionalized heteroaromatic compounds successfully (Scheme 5.1) [17, 18, 19, 20].

The palladium catalyzed Heck reaction of 2-pyridyl tosylates would not only provide access to the α-arylated olefins but also act as direct precursors to the attractive chiral benzylic amine and amide derivatives often found in bioactive compounds.

5.3 Results and Discussion

The pyridyl tosylates needed for this study were obtained in good yields by simple tosylation of the commercially available 2-hydroxypyridines using tosyl chloride in the presence of base and the organic catalyst *N,N*-dimethylaminopyridine (DMAP). However, *N*-tosylation were in some cases observed, due to the intrinsic tautomerization of the 2-hydroxypyridine. Based on previous reported catalytic systems for the regioselective Heck reaction of electron rich olefins, promising results were obtained applying $Pd(dba)_2$ and DPPF in the olefination of 2-pyridyl tosylate **47** using *N*-vinylacetamide (NVA) in the combination of the tertiary base Cy_2NMe in dioxane at 85 °C. This afforded the α-substituted alkene **48** with perfect regioselectivity (Table 5.1, entry 1) [6].

Gratifyingly, full conversion was attained after 24 h when increasing the reaction temperature to 100 °C, and an 87 % yield could be secured upon column chromatography (entry 2). Changing to the tertiary amine base diisopropylethylamine (DIPEA) also led to full conversion but only resulted in a 66 % isolated yield (entry 3). Other bisphosphine ligands such as DPPP or (±)-BINAP proved inferior compared to DPPF (entries 4 and 5). In addition, DPPE, DPPPe (1,2-bis(diphenylphosphino)pentane) and D*i*PrPF resulted in incomplete reaction but still afforded the α-arylated olefin exclusively, demonstrating the generation of the intermediary cationic palladium(II) species to be rather flexible (results not

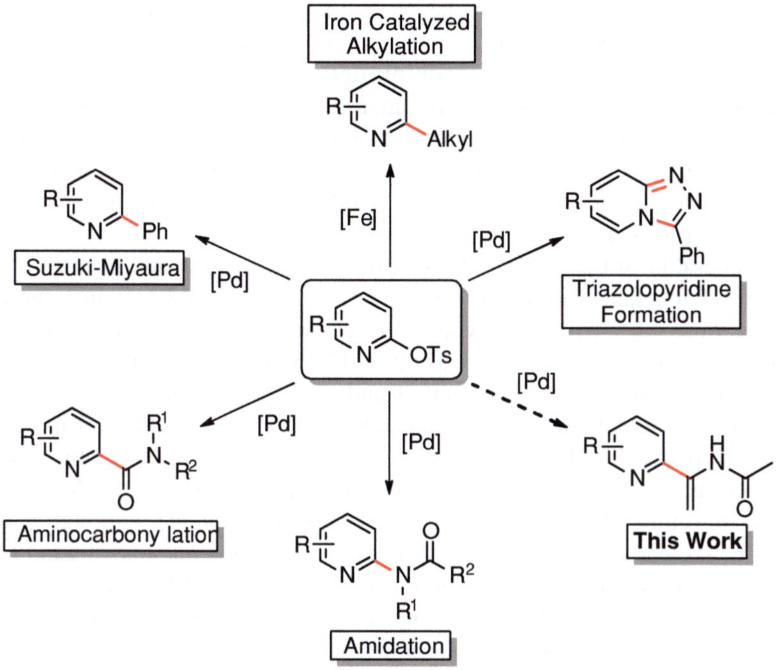

Scheme 5.1 Transition metal catalyzed reactions utilizing pyridyl tosylates

Table 5.1 Optimization of the heck reaction employing pyridyl tosylates

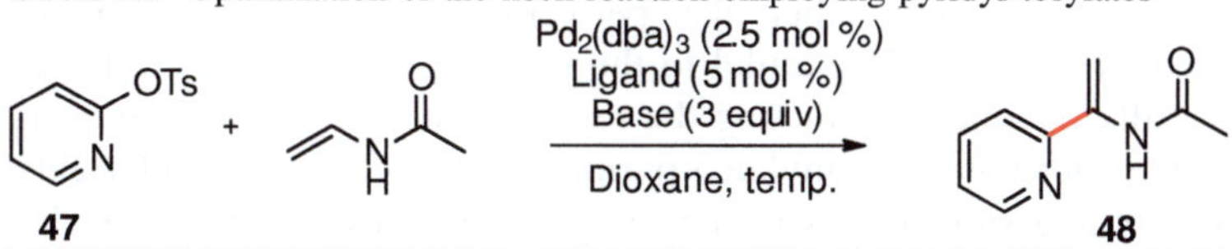

Entry[a]	Ligand	Base	T (°C)	Yield (%)[b] (conversion)
1	DPPF	Cy_2NMe	85	(63)
2	DPPF	Cy_2NMe	100	87 (100)
3	DPPF	DIPEA	100	66 (100)
4	DPPP	Cy_2NMe	100	78 (100)
5	BINAP	Cy_2NMe	100	74 (100)
6	DPPF	Cy_2NMe	70	(50)[c]
7	DPPF	Cy_2NMe	100	71 (100)[d]
8	DPPF	Cy_2NMe	100	72 (100)[e]

[a] **47** (1 equiv), NVA (4 equiv). [b] Yield of the isolated product. [c] THF was used as the solvent. [d] **47** (1.5 equiv), NVA (1 equiv). [e] 2-Pyridyl mesylate (**49**) was used instead of **47** Reprinted (adapted) with permission from *Chem. Eur. J.* **2009**, *15*, 5950. Copyright 2009 Wiley–VCH Verlag GmbH & Co. KGa. Weinheim

shown). However, mixtures of the α-, and β-isomers were observed, when employing 1,10-phenanthroline as the auxiliary ligand, potentially explained by the weaker coordinating abilities of nitrogen based ligands compared to the phosphino derived analogous. Consequently, the competing neutral pathway would thereby influence the regioselective outcome (results not shown). Conducting the reaction in THF or altering the ration of **47** and NVA did not improve the catalytic turn-over (entries 6 and 7). Additionally, attempts to lower the catalytic loading turned out unsuccessful (results not shown).

The generality of the optimized conditions was then evaluated against a variety of 2-pyridyl tosylates holding different substituents (Table 5.2) In general, the Heck reactions of 2-pyridyl tosylate derivatives and *N*-vinylacetamide went to completion after 22 h resulting in good yields. Functional groups encompassing esters, amides, chlorides, trifluormethyl and nitro groups were allowed, complementing the excellent functional group tolerance typically observed in the Heck reaction. In particular, the presence of a chloride unravels the possibility for further functionalization via transition metal catalysis (entries 1, 2, 6 and 9). As depicted in Table 5.2 both DPPP and DPPF proved adaptable, although a slightly more effective catalyst was generated using DPPF as the ancillary ligand (entries 1, 6, 9, 10). However, the introduction of a nitro group lowered the catalytic outcome considerably, and a moderate 25 % yield was secured (entry 8). A satisfactory 45 % yield was attained in the double coupling of the 2,6-pyridyl ditosylate, although incomplete reaction was detected (entry 12). Interestingly, when the ditosylated 2,4-dihydroxyquinoline was employed, the product originating from a double olefination was produced in a 55 % yield, suggesting 4-pyridyl tosylates to be viable substrates as well (entry 13).

To our delight, *n*-butyl vinyl ether also performed well under the applied reaction conditions (Table 5.3). In parallel to the reactions using NVA, full conversion was generally obtained conducting the reactions at 100 °C for 22 h using either DPPF or DPPP forming the vinylated pyridines in good to excellent yields. Pyridyl tosylates holding substituents in the three-position resulted in somewhat lower yields (entries 2 and 3). In contrast to the double olefinated quinoline provided in the Heck reaction of NVA, the mono-vinylated 4-tosylquinoline was obtained in 60 % yield suggesting, that NVA is considerably more reactive than *n*-butyl vinyl ether (entry 5). This was confirmed by a competition experiment between NVA and *n*-butyl vinyl ether forming the α-arylated NVA exclusively (results not shown). On the other hand, the nitro group containing pyridyl tosylate performed better when coupled to butyl vinyl ether producing the 1,1-disubstituted alkene in an acceptable 46 % isolated yield (entry 6). Introduction of standard protecting groups such as Cbz and Boc on both the nucleophilic and electrophilic coupling partner were well tolerated. Olefination of 2-pyridyl tosylate carrying a Cbz-protected β-amino ester could be secured in a 59 % isolated yield (Table 5.3, entry 10). Also, the Boc protected enamide coupled smoothly to 2-pyridyl tosylates holding a trifluoromethyl and chloride group yielding **50** and **51** in 57 and 55 %, respectively (Scheme 5.2).

Table 5.2 Regioselective heck reaction utilizing *N*-vinylacetamide

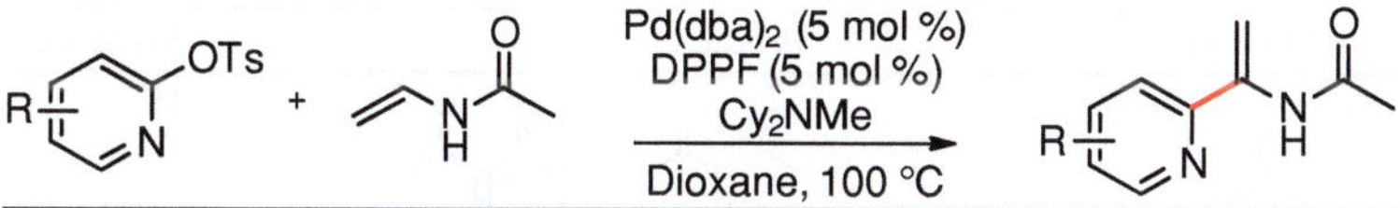

Entry[a]	Aryl tosylate	Product	Yield (%)[b]
1	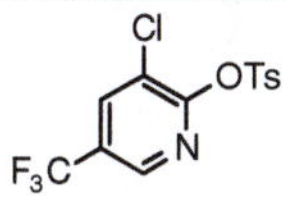	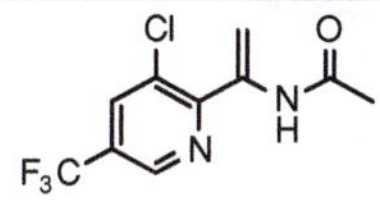	68 (64)[c]
2	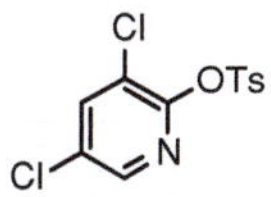	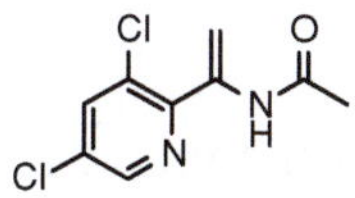	68[d]
3	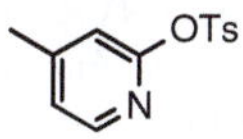	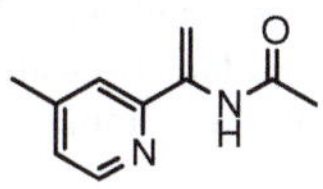	74
4	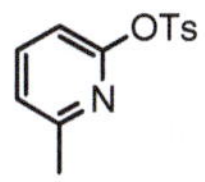	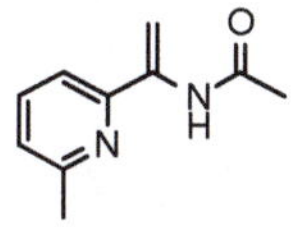	81
5	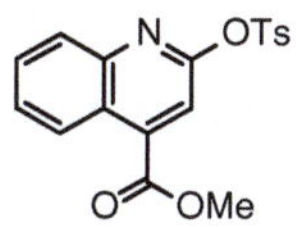	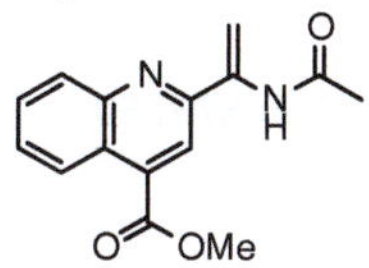	74
6	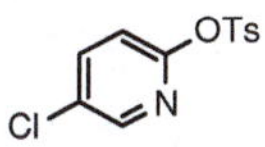	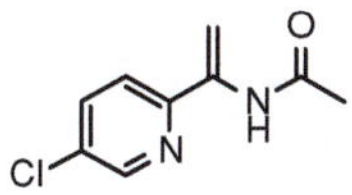	88 (72)[c]
7	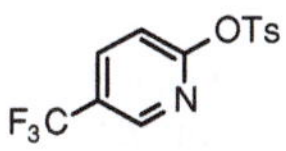	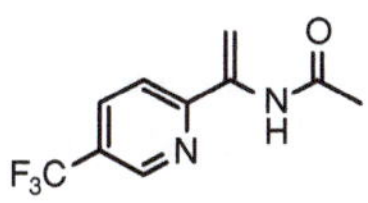	(81)[c]
8	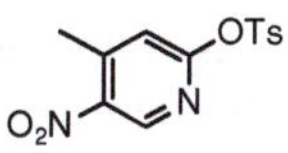	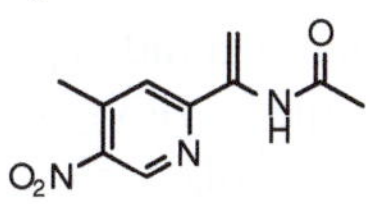	(25)[c]
9	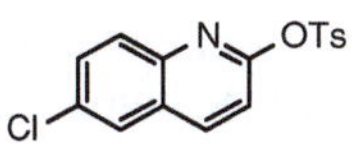	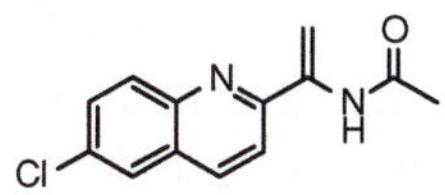	79 (74)[c]
10	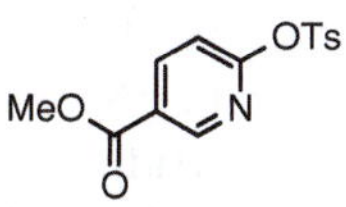	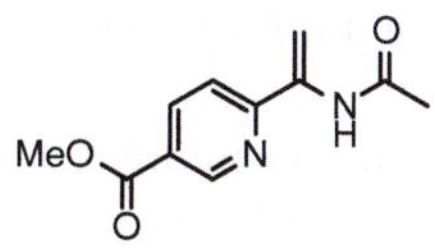	73 (55)[c]
11	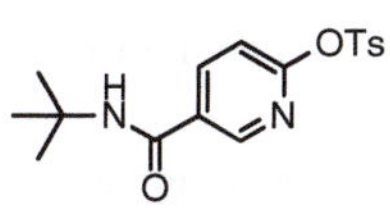	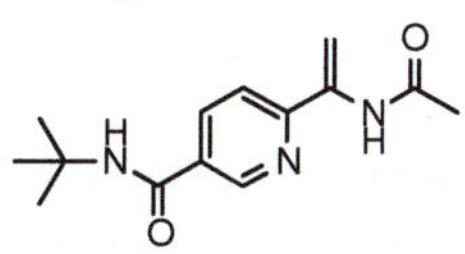	75

(continued)

Table 5.2 (continued)

Entry[a]	Aryl tosylate	Product	Yield (%)[b]
12	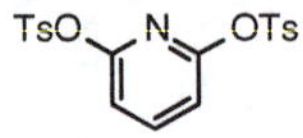	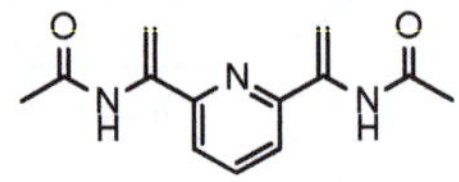	45
13	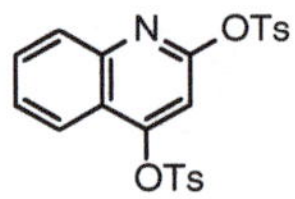	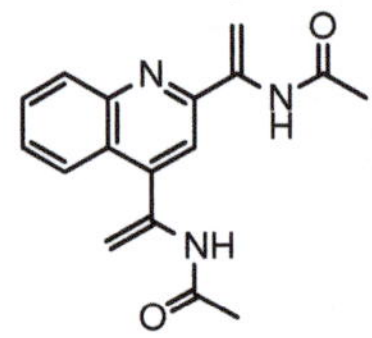	(55)[c]

[a] 2-Pyridyl tosylate (0.3 mmol), NVA (1.2 mmol), Cy_2NMe (0.9 mmol). [b] Yield of the isolated product. [c] Yields in brackets obtained with DPPP as the ligand. [d] NVA (1 equiv) 2-pyridyl tosylate (1.5 equiv) Reprinted (adapted) with permission from *Chem. Eur. J.* 2009, *15*, 5950. Copyright 2009 Wiley–VCH Verlag GmbH & Co. KGa. Weinheim

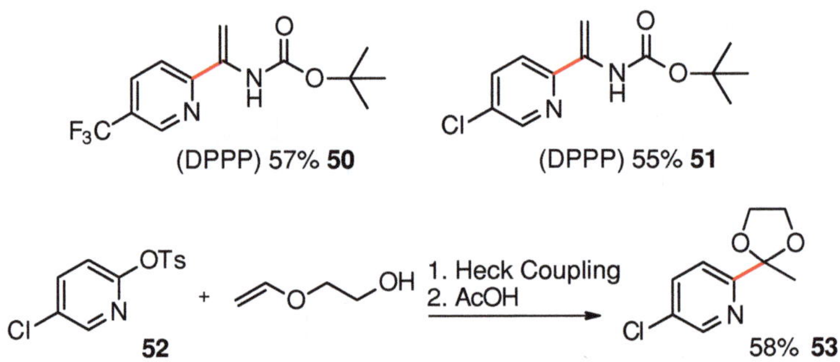

Scheme 5.2 Heck reactions of functionalized olefins. Reprinted (adapted) with permission from *Chem. Eur. J.* **2009**, *15*, 5950. Copyright 2009 Wiley–VCH Verlag GmbH & Co. KGa. Weinheim

Cyclic and *N*-alkylated enamides proved ineffective as coupling partners in the Heck reaction forming the pyridyl olefins only to a limited extent. In contrast, olefins containing free alcohols performed well as illustrated by glycol vinyl ether obtaining the cyclized ketal **53** after acetic acid treatment of the resulting α-arylated *n*-butyl vinyl ether in a 58 % overall yield (Scheme 5.2).

Next, tosylated hydroxypyrimidines were tested in the regioselective Heck reaction using NVA and *n*-butyl vinyl ether in order to investigate the influence of the heteroaromatic core system on the catalytic outcome (Scheme 5.3). Similar yields of the vinylated pyrimidines compared to the pyridine equivalents were achieved indicating a robust catalytic system. Even a double Heck reaction could be accomplished yielding **56** in 36 % corresponding to an average of 60 % yield of each vinylation.

In order to reveal the full potential of the optimized reaction conditions, it was decided to examine the scale-up applications of the catalytic system (Scheme 5.4). 5-Chloro-2-pyridyl tosylate **52** was chosen as test substrate based on its performances in the initial investigations (Table 5.2, entry 6 and Table 5.3, entry 7).

Table 5.3 Regioselective heck reaction utilizing *n*-butyl vinyl ether

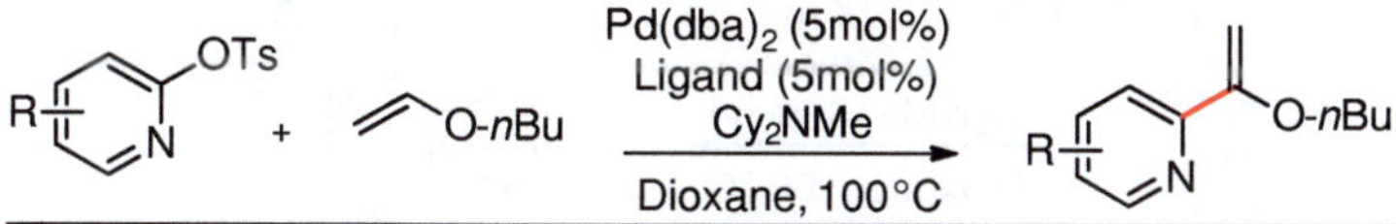

Entry[a]	Aryl tosylate	Product	Ligand	Yield (%)[b]
1	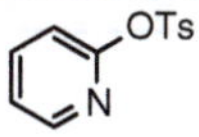	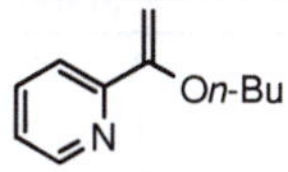	DPPF	81
2	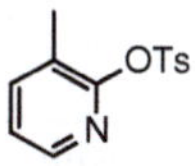	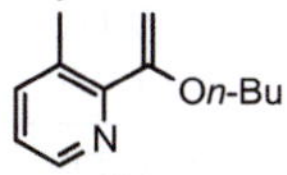	DPPF	49
3	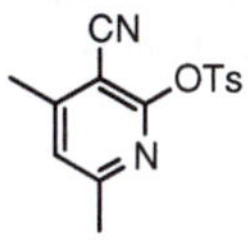	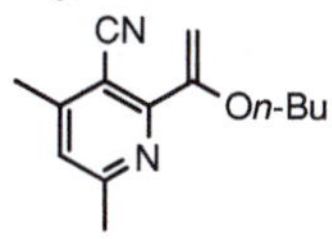	DPPP	69
4	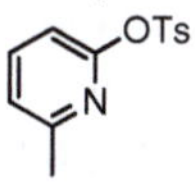	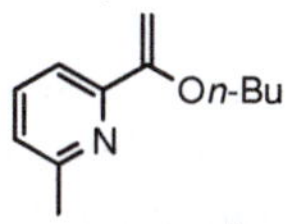	DPPF	86
5	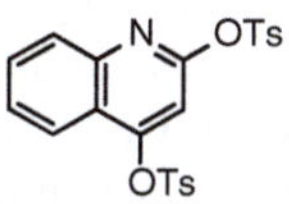	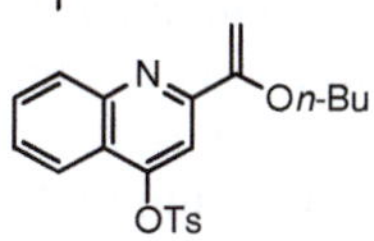	DPPP	60
6	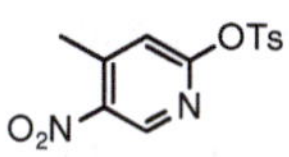	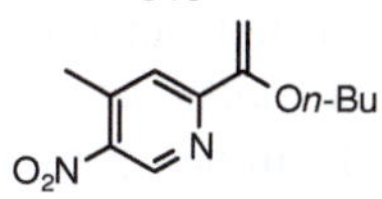	DPPF	46
7	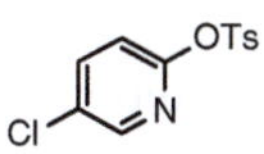	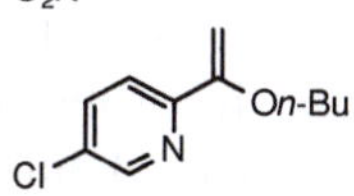	DPPP	81
8	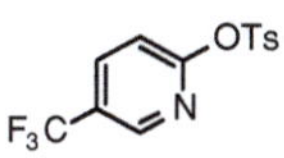	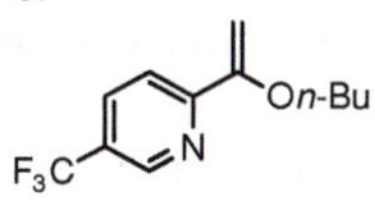	DPPP	78
9	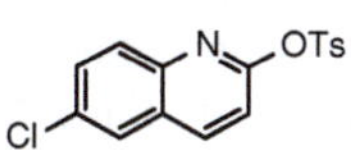	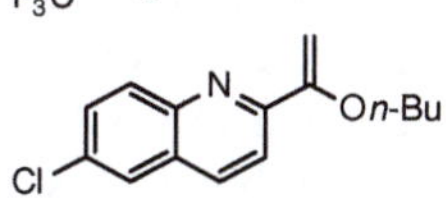	DPPP	82
10	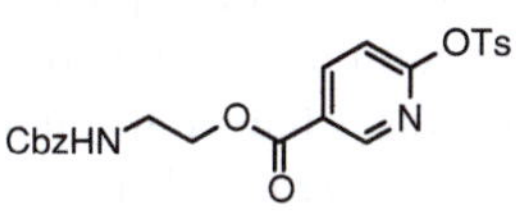	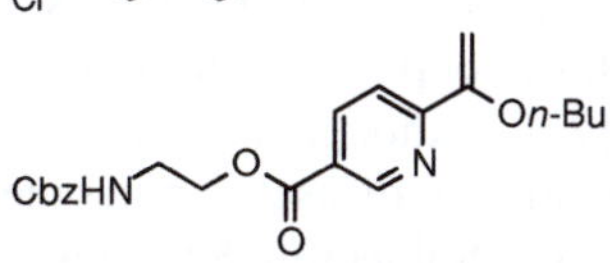	DPPF	59

[a] 2-Pyridyl tosylate (0.3 mmol), *n*-butyl vinyl ether (1.2 mmol), Cy_2NMe (0.9 mmol) [b] Yield of the isolated product Reprinted (adapted) with permission from *Chem. Eur. J.* **2009**, *15*, 5950. Copyright 2009 Wiley–VCH Verlag GmbH & Co. KGa. Weinheim

Conducting the reaction on a scale 10 times greater than the standard reaction and lowering the catalytic loading from 5 to 3 mol% resulted in an 89 % yield (entry 1). Lower amounts of dioxane were applied relative to the standard reactions leading

Table 5.4 Scale-Up applications of the heck reaction

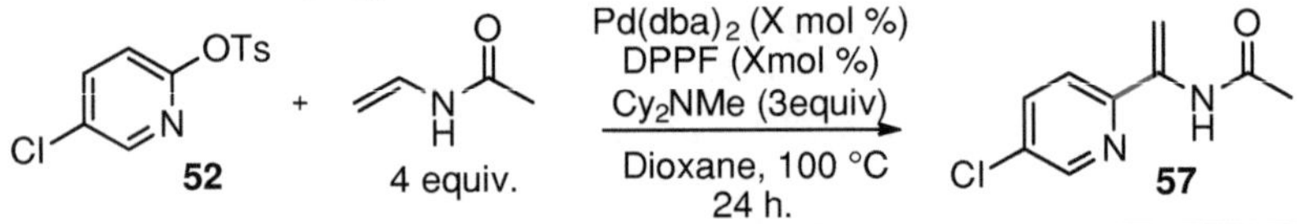

Entry	Scale (mmol)	Catalytic loading (%)	Yield (%)[a]
1	3	3	89
2	3	1	92
3	3	0.1	61[b]
4	15	1	88

[a] Yield of the isolated product. [b] 64 % conversion after 48 h.Reprinted (adapted) with permission from *Chem. Eur. J.* **2009**, *15*, 5950. Copyright 2009 Wiley–VCH Verlag GmbH & Co. KGa. Weinheim

to higher concentrations of the catalyst demonstrating increased stability of the catalytic system, when operating on a larger scale. Reducing the catalytic loading to 1 mol% improved the turn over number, and a 92 % isolated yield of the vinylated chloropyridine could be secured upon column chromatography (entry 2). However, prolonged reaction time was required, when 0.1 mol% of the DPPF derived catalyst was employed providing the desired product in 61 % (entry 3). Finally, a comparable yield corresponding to 2.59 g was acquired, when the Heck reaction of **52** and NVA was carried out on a scale 50 times greater than the standard reaction applying only 1 mol% catalyst (entry 4).

In general, the pyridyl vinyl amides proved more stable than the products originating from *n*-butyl vinyl ether in accordance to the literature, affording the corresponding aryl methyl ketone when subjected to acidic conditions [21, 22]. Thus, the capability of the α-arylated vinyl amides possessing additional reactive site to undergo further functionalization was assessed using **52** as test substrate (Scheme 5.4).

Treating **52** with aqueous hydrochloric acid afforded the pyridyl methyl ketone **58** in 93 %, whereas the dechlorinated benzylic amide **59** was obtained in near quantitative yields after reduction using palladium on charcoal (Path **A** and **B**) [23, 24]. Activation of the C–Cl bond was successfully accomplished in the copper-free Sonogashira coupling, using reaction conditions developed by Buchwald and coworkers [25]. Although a moderate yield was achieved using *tert*-butyl acetylene, a synthetically more useful outcome was secured with the TIPS-protected acetylene (Path **C**). Subjecting **52** to an additional Heck reaction employing reaction conditions extrapolated from the conditions reported Fu and coworkers provided the double vinylated pyridine using either acrylamide or acrylate in 63 and 72 %, respectively (Path **D**) [26]. Finally, a biaryl scaffold was effectively constructed through functionalization of the pyridyl chloride by a palladium catalyzed Suzuki coupling (Path **E**) [27]. These results clearly demonstrate the readily accessible pyridyl tosylates as viable building blocks in the synthesis of highly functionalized heteroaromatic compounds.

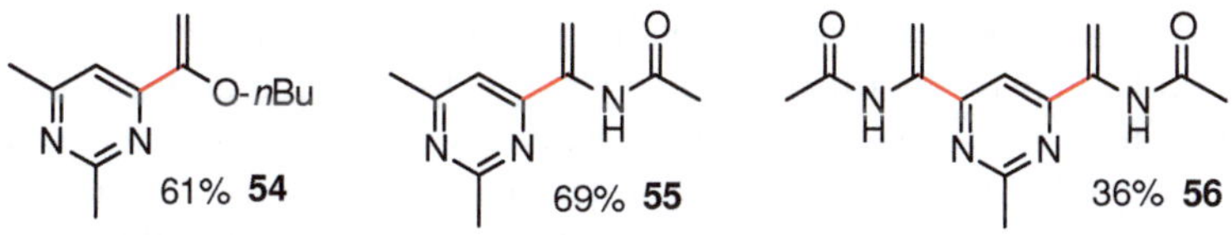

Scheme 5.3 Pyrimidyl tosylates in the regioselective heck reaction. Reprinted (adapted) with permission from *Chem. Eur. J.* **2009**, *15*, 5950. Copyright 2009 Wiley–VCH Verlag GmbH & Co. KGa. Weinheim

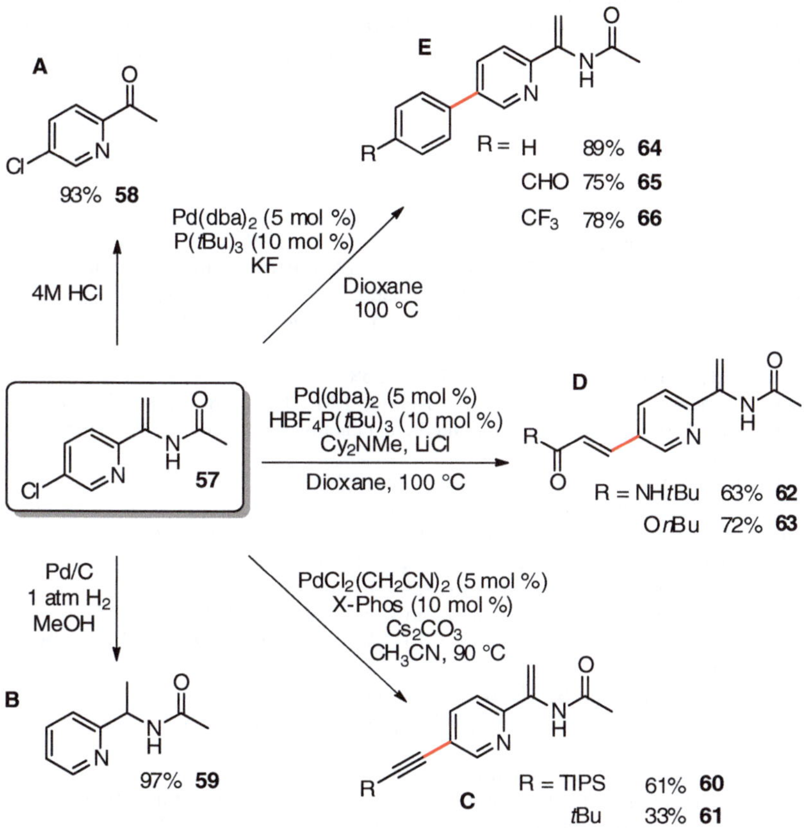

Scheme 5.4 Functionalization of *N*-(1-(5-chloropyridin-2-yl)vinyl)acetamide **57**. Reprinted (adapted) with permission from *Chem. Eur. J.* **2009**, *15*, 5950. Copyright 2009 Wiley–VCH Verlag GmbH & Co. KGa. Weinheim

5.4 Conclusion and Outlook

In conclusion, the first general protocol describing the palladium catalyzed Heck reaction of heteroaromatic tosylates and electron rich olefins has been established based on a catalyst derived from conventional bisphosphine ligands. Furthermore, the high levels of regioselectivity obtained, confirm the formation of the crucial cationic palladium(II) intermediate generated by dissociation of the tosylate anion.

In order to evaluate the generality of the catalytic system, a variety of different pyridyl and pyrimidyl tosylates were tested and proved effective facilitating the α-arylation of *N*-vinylacetamide and *n*-butyl vinyl ether successfully. In particular, scale-up experiments confirmed the high efficiency and pronounced stability of the applied palladium catalyst relying on DPPF as the ligand. The high levels of stability displayed by the α-arylated *N*-vinylacetamide allowed for the further functionalization via palladium catalyzed transformations.

Further developments of the regioselective Heck reaction between aryl sulfonates and electron rich olefins will be presented in the following chapter.

References

1. Nguyen HN, Huang X, Buchwald SL (2003) J Am Chem Soc 125:11818
2. Choy PY, Chow WK, So CM, Lau CP, Kwong FY (2010) Chem Eur J 16:9982
3. Ackermann L, Kapdi AR, Fenner S, Kornhaaß C, Schulzke C (2011) Chem Eur J 17:2965
4. Hesp KD, Lundgren RJ, Stradiotto M (2011) J Am Chem Soc 133:5194
5. Ogata T, Hartwig JF (2008) J Am Chem Soc 130:13848
6. Hansen AL, Skrydstrup T (2005) Org Lett 7:5585
7. Slagt VF, de Vries AHM, de Vries JG, Kellogg RM (2010) Org Process Res Dev 14:30
8. Patrick GL (2005) An introduction to medicinal chemistry, 3rd edn. Oxford University Press, New York
9. Sha F, Huang X (2009) Angew Chem Int Ed 48:3458
10. Hill MD, Movassaghi M (2008) Chem Eur J 14:6836
11. Ohta Y, Tokimizu Y, Oishi S, Fujii N, Ohno H (2010) Org Lett 12:3963
12. Neumann JJ, Suri M, Glorius F (2010) Angew Chem Int Ed 49:7790
13. Campeau L-C, Rousseaux S, Fagnou K (2005) J Am Chem Soc 127:18020
14. Cho SH, Hwang SJ, Chang S (2008) J Am Chem Soc 130:9254
15. Peréz-Balado C, Willemsens A, Ormerod D, Aelterman W, Mertens N (2007) Org Process Res Dev 11:237
16. Billingsley K, Buchwald SL (2007) J Am Chem Soc 129:3358
17. Hermange P, Lindhardt AT, Taaning RH, Bjerglund K, Lupp D, Skrydstrup T (2011) J Am Chem Soc 133:6061
18. Gøgsig TM, Lindhardt AT, Skrydstrup T (2009) Org Lett 11:4886
19. Mantel MLH, Lindhardt AT, Lupp D, Skrydstrup T (2010) Chem Eur J 16:5437
20. Lindhardt AT, Gøgsig TM, Gauthier D, Lupp D, Mantel MLH, Bjerglund KM, Skrydstrup T (2010) Isr J Chem 50:558
21. Liu Z, Xu D, Tang W, Xu L, Mo J, Xiao J (2008) Tetrahedron Lett 49:2756
22. Andappan MMS, Nilsson P, von Schenck H, Larhed M (2004) J Org Chem 69:5212
23. Hansen AL, Skrydstrup T (2005) J Org Chem 70:5997
24. Harrison P, Meek G (2004) Tetrahedron Lett 45:9277
25. Gelman D, Buchwald SL (2003) Angew Chem Int Ed 42:5993
26. Littke AF, Fu GC (2001) J Am Chem Soc 123:6989
27. Littke AF, Dai C, Fu GC (2000) J Am Chem Soc 122:4020

Chapter 6
Cationic Intermediates in the Ni(0)-Catalyzed Heck Reaction

6.1 Introduction

While palladium based complexes are the most common catalysts for the Heck reaction, efforts have been made to replace palladium with less expensive metals such as nickel or cobalt [1]. During the last decade increasing attention has been directed towards nickel catalysis. The sustainability and relatively low cost of nickel precursors are key factors contributing to the attractiveness when considering transition metal catalysis. Numerous reports have been disclosed on a wide variety of different nickel catalyzed processes including cross couplings, [2, 3] cycloisomerization, [4, 5] cycloadditions, [6, 7] multicomponent couplings, [8, 9] annulations, [10, 11] reductive couplings, [12, 13] etc. [14, 15]. Perhaps, one of the most striking features of nickel catalysis compared to palladium is the enhanced range of adaptable electrophiles. In addition to aryl and vinyl halides and sulfonates, functional groups traditionally inert towards oxidative addition such as ethers, [16] metal alkoxides, [17] carboxylates, [18] carbamates [19] and so forth have been accessed and utilized in a variety of cross coupling reactions. Especially the work from the groups of Shi, Garg and Chatani has paved the way for these classes of C–O compounds as viable coupling partners [20, 21]. This in turn allows for the sequential functionalization of aromatic systems containing multiple reactive sites demonstrating the flexibility displayed by nickel catalysis.

Parts of the following chapter have been published in *J. Am. Chem. Soc.* **2012**, *134*, 443 and have been reprinted (adapted) with permission from *J. Am. Chem. Soc.* **2012**, *134*, 443.

T. M. Gøgsig, *New Discoveries on the β-Hydride Elimination*, Springer Theses,
DOI: 10.1007/978-3-642-32099-6_6, © Springer-Verlag Berlin Heidelberg 2012

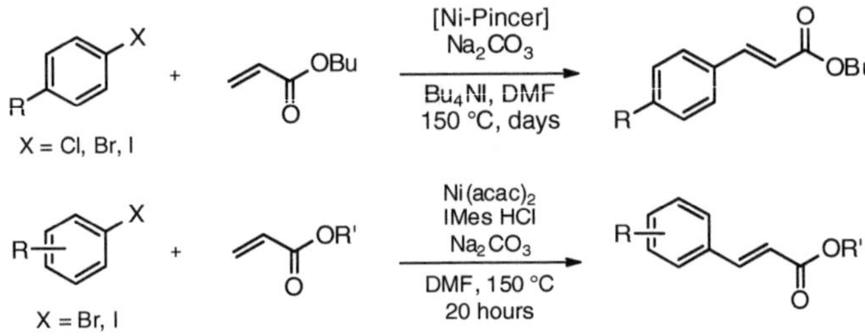

Scheme 6.1 Nickel catalyzed Heck reaction. Reprinted (adapted) with permission from *J. Am. Chem. Soc.* **2012**, *134*, 443. Copyright 2011 American Chemical Society

6.2 Background

Alkyl halides have been employed as electrophiles with great success in nickel catalysis creating both C(sp^3)—C(sp^3) and C(sp^3)—C(sp^2) σ-bonds under mild reaction conditions [22]. In particular, enantioselective protocols relying on a nickel catalyst carrying chiral ligands have been developed by Fu and coworkers [23]. Furthermore, alkyl halides possessing β-hydrogens have proven perfectly suitable and products arising from the undesired β-hydride elimination are usually not observed. This could be accounted for by an inherent reluctance of nickel to undergo β-hydride eliminations, although procedures relying on this transformation are well established including the Shell higher olefin process [24, 25]. Nevertheless, in contrast to other cross coupling reactions, nickel catalyzed Heck couplings have only been scarcely described in the literature. Harsh reaction conditions such as high temperatures, high polar solvents and prolonged reaction times are usually required, for these reactions to proceed (Scheme 6.1) [26, 27].

This could indicate that the β-hydride elimination generating the nickel hydride intermediate, crucial to the Heck reaction, is in fact problematic. An alternative explanation for the observed difficulties in the nickel catalyzed Heck reactions could also be the stronger bond energy of the Ni–H compared to the Pd–H complex resulting in a more difficult reductive elimination. DFT calculations performed by Guo and coworkers indicated, that the β-hydride elimination from nickel(II) complexes is indeed slower compared to the palladium analogue displaying a free energy difference of approximately 30 kcal mol^{-1}. In addition, the reductive elimination from nickel(II) hydride complexes was found to be considerably more difficult than from the corresponding palladium(II) hydride complexes, revealing an endergonic process (16 kcal mol^{-1}) for the nickel based reductive elimination. On the other hand, the corresponding palladium catalyst regeneration was calculated to be exergonic with a free energy of −4 kcal mol^{-1} [28]. Hence, certain properties of both the base and the catalytic system would be needed in order to regenerate the catalyst and complete the catalytic cycle.

Jamison and coworkers recently reported on a protocol describing the nickel catalyzed vinylation of allylic ethers and carbonates producing 1,4-dienes at

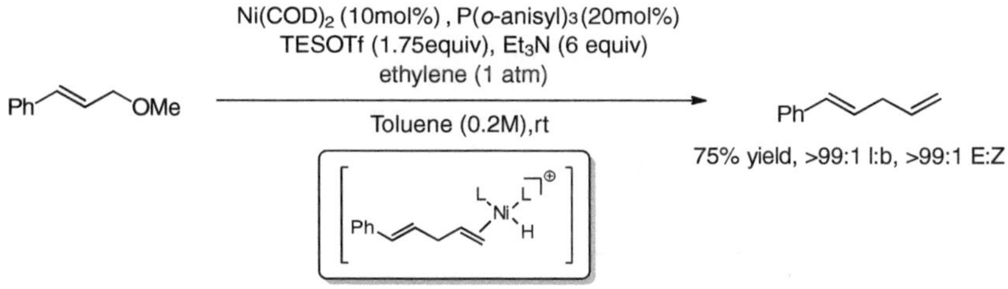

Scheme 6.2 Nickel catalyzed allylic substitutions of simple alkenes. Reprinted (adapted) with permission from *J. Am. Chem. Soc.* **2012**, *134*, 443. Copyright 2011 American Chemical Society

ambient temperature (Scheme 6.2) [29]. The underlying mechanism was proposed to proceed through a β-hydride elimination from a cationic nickel(II) complex forming a cationic hydridonickel(II) complex. Subsequent base induced reductive elimination would then regenerate the catalyst. TESOTf was added to the reaction mixture in order to secure the formation of the cationic nickel(II) intermediate, abstracting the alkoxy group from the nickel nucleus leaving the poor coordinating triflate counterion behind. Noteworthy, the reaction proceeded smoothly without the needs of high polar solvents such as ionic liquids or high temperatures, which could imply, that the cationic nickel(II) intermediate is of utmost importance.

In the light of the results presented by the group of Jamison and our own discoveries regarding the palladium catalyzed reaction between aryl and vinyl tosylates and electron rich olefins vide supra, it was decided to investigate the nickel catalyzed arylation of electron rich olefins in more detail [30].

6.3 Results and Discussion

As described in Sect. 1.7.1 the regioselective Heck reaction, producing the α-arylated olefins selectively, proceeds by a pathway involving a vital cationic palladium(II) intermediate. This pathway can be favored by a number of factors, including triflate based electrophiles and bidentate ligands (see Sect. 1.7.1).

A catalytic system derived from $Ni(COD)_2$ and a bidentate phosphine ligand in combination with the tertiary amine base Cy_2NMe was envisioned to be a suitable starting point in the coupling of 4-biphenyl triflate **67** to butyl vinyl ether. Preliminary, screenings revealed, that this was indeed possible (Table 6.1). Conducting the reaction in dioxane at 100 °C for approximately 20 h resulted in a 21 % conversion using DPPE as the ligand (entry 1). Increasing the bite angle of the bisphosphine facilitated the desired reaction to a greater extent (entries 2 and 3) [31].

Gratifyingly, full conversion was obtained, when DPPF was employed, and a 90 % isolated yield of 4-phenylacetophenone **68** could be secured after acidic hydrolysis of the initially formed α-arylated butyl vinyl ether (entry 4). Catalytic systems carrying more bulky and electron donating bidentate ligands did not provide

Table 6.1 Optimization of the nickel catalyzed heck reaction

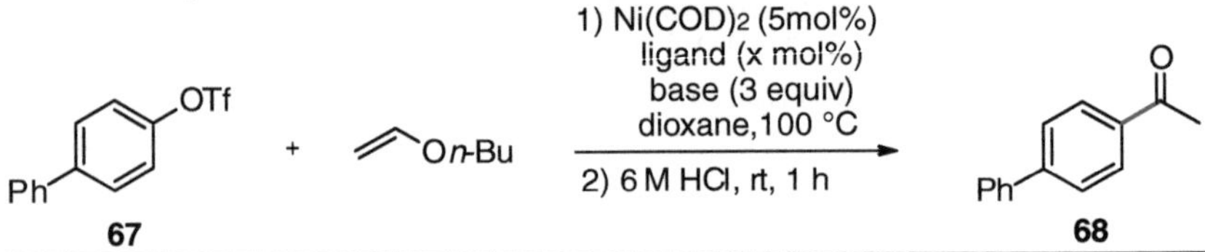

Entry[a]	Ligand (mol%)	Base	Conversion (%)[b]
1	DPPE (5)	Cy_2NMe	21
2	DPPP (5)	Cy_2NMe	46
3	BINAP (5)	Cy_2NMe	42
4	DPPF (5)	Cy_2NMe	100 (90)
5	PPF-*t*-Bu (5)	Cy_2NMe	<5
6	D*i*PrPF (5)	Cy_2NMe	<5
7	D*t*BuPF (5)	Cy_2NMe	<5
8	Xantphos (5)	Cy_2NMe	<5
9	PCy_3HBF_4 (10)	Cy_2NMe	<5
10	PPh_3 (10)	Cy_2NMe	<5
11	DPPF (5)	DIPEA	76
12	DPPF (5)	Et_3N	69
13	DPPF (5)	Pyridine	<5
14	DPPF (5)	DBU	<5

[a] Aryl triflate (0.15 mmol), butyl vinyl ether (0.6 mmol), base (0.45 mmol), dioxane 1 mL at 100 °C for 20 h. [b] Based on ^{1}H NMR analysis. Isolated yields in parenthesis.
Reprinted (adapted) with permission from *J. Am. Chem. Soc.* **2012**, *134*, 443. Copyright 2011 American Chemical Society

any detectable amounts of the desired product (entries 5–8). Applying monodentate phosphine ligands, which have demonstrated great performances in other cross couplings, turned out to be nonproductive (entries 9 and 10) [2, 3]. Different amine bases were then evaluated. Diisopropylethylamine (DIPEA) and triethylamine, often used in palladium catalyzed Heck reaction, were inferior compared to Cy_2NMe (entries 11 and 12). On the other hand, no reaction was observed when changing the base to pyridine or increasing the basicity as in the case of DBU (entries 13 and 14). Finally, a significantly drop in the reaction turnover was observed when lowering the reaction temperature to 80 °C, whereas changing the catalyst precursor to $Ni(acac)_2$ resulted in no reaction (results not shown).

With these optimized conditions in hand we set forth to investigate the generality of this reaction. A variety of aryl triflates were synthesized using standard procedures and evaluated against butyl vinyl ether (Table 6.2) [32]. In general, good to excellent yields of the desired aryl methyl ketones were obtained after treating the resulting aryl vinyl ether with 6 M HCl at room temperature for approximately 1 h. Aryl triflates containing electron withdrawing or donating groups performed well under the applied reaction conditions. Functionalities such as esters, amides, cyano, diazo, acetals and trifluoromethyl groups were perfectly tolerated, paralleling the functional group tolerance usually observed in the

Table 6.2 Nickel catalyzed heck reaction of aryl triflates

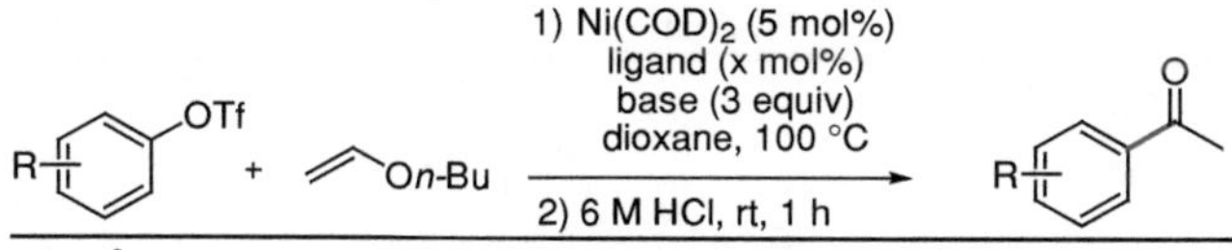

Entry[a]	Aryl triflate	Product	Yield (%)[b]
1	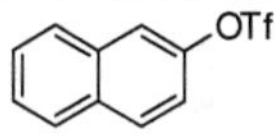	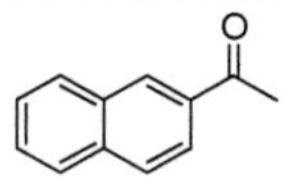	91
2	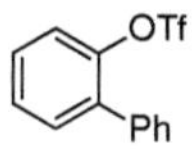	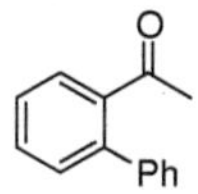	88[c]
3	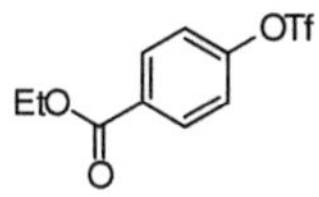	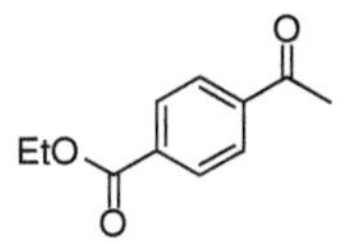	79
4	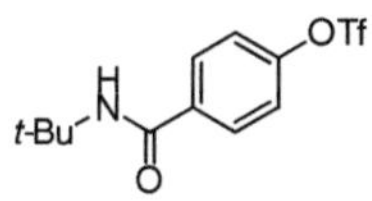	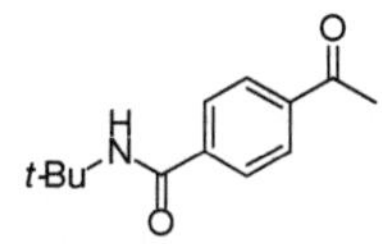	90
5	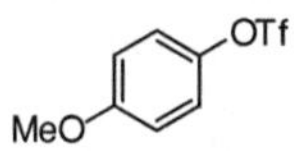	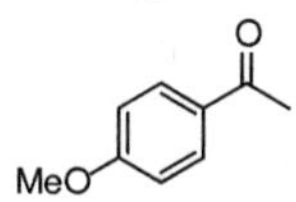	93
6	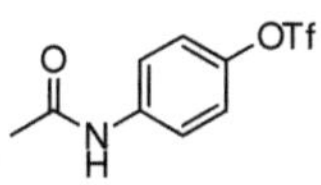	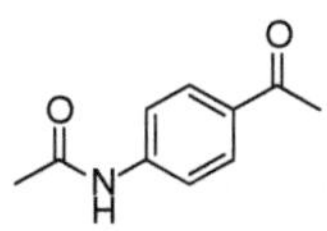	56
7	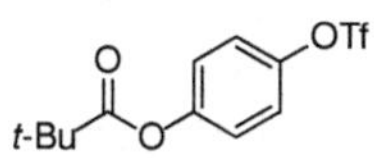	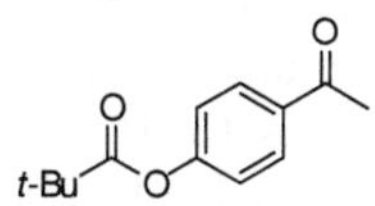	83[c]
8	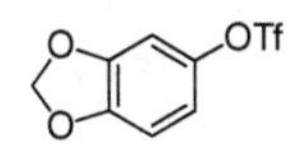	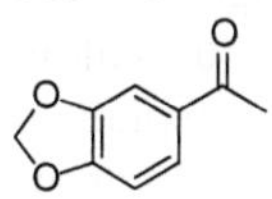	77[c]
9	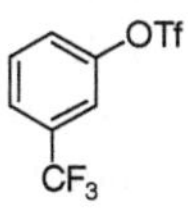	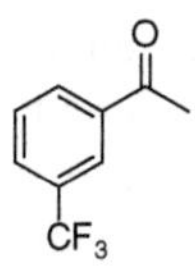	81
10	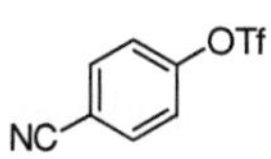	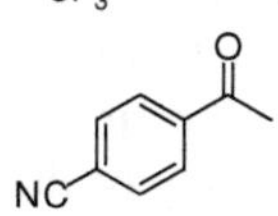	78
11	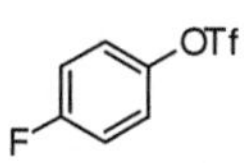	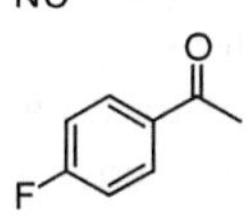	78[c]

(continued)

Table 6.2 (continued)

Entry[a]	Aryl triflate	Product	Yield (%)[b]
12	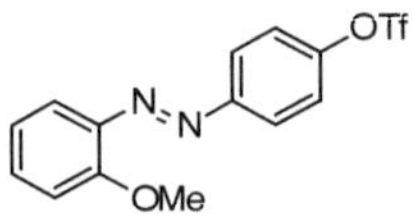	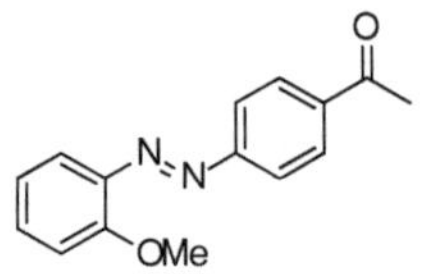	75
13	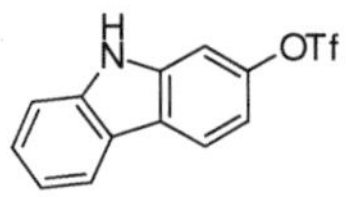	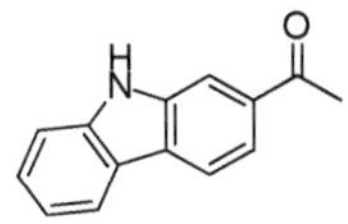	76
14	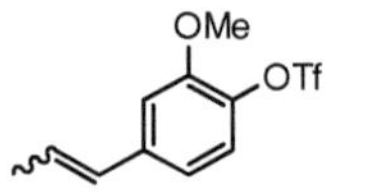	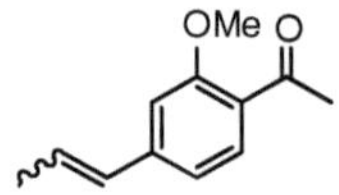	97
15	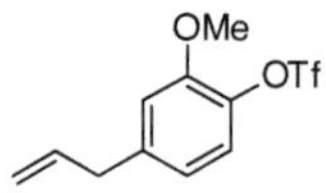	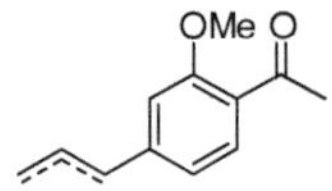	92[d]
16	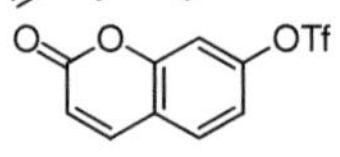	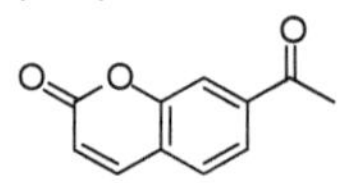	78

[a] Aryl triflate (0.5 mmol), butyl vinyl ether (2.0 mmol), Cy_2NMe (1.5 mmol), DPPF (5 mol%), $Ni(COD)_2$ (5 mol%) in dioxane 3 mL at 100 °C for 20 h. [b] Isolated yields. [c] DPPF (7 mol%) and $Ni(COD)_2$ (7 mol%). [d] A 7:10:2 mixture (*E*:*Z*:allyl) was obtained

analogous palladium catalyzed Heck reaction (entries 3–12). Furthermore, acetylation of biological relevant structures represented by carbazole, [33] eugenol, [34] isoeugenol [34] and coumarin [35] were successfully achieved attaining near quantitative yields (entries 13–16). Interestingly, isomerization of the allyl double bond in the coupling of eugenol triflate was identified, generating mixtures of the conjugated *cis* and *trans* adducts along with the non-isomerized product (entry 15). Potentially, this could be explained by a long lived nickel hydride complex responsible for a consecutive hydronickelation and β-hydride elimination forming the thermodynamically more stable styrene derivative [36]. Somewhat lower yields were in some cases obtained, when electron donating groups were positioned *para* to the triflate or *ortho* substituents were introduced on the aromatic ring (entries 2, 6–8). However, a slight increase in the catalytic loading restored the high levels of the synthetically useful outcome.

Having established an efficient catalytic system for the nickel catalyzed Heck reaction, a more in-depth study of the mechanistic details was desired. In order to investigate the role of the leaving group, aryl sulfonates with different electronic properties were prepared from 2-naphthol and various phenylsulfonyl chloride derivatives, using the same procedures as described in Chap. 5 (Table 6.3). Relying on the optimized reaction conditions, the coupling between 2-naphthyl benzenesulfonate and *n*-butyl vinyl ether resulted in a 68 % conversion (entry 1). Mixtures of

Table 6.3 Effect of the leaving group on the nickel catalyzed heck reaction

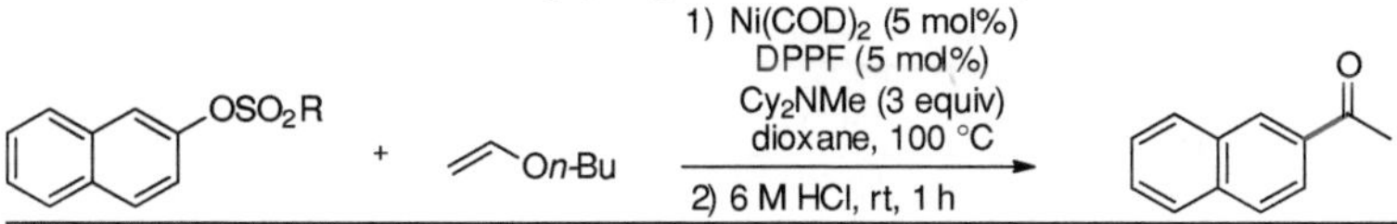

Entry[a]	R	Conversion (%)[b]
1	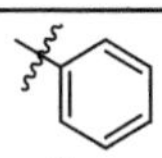	68[c]
2	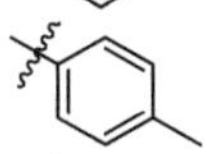	40[c]
3	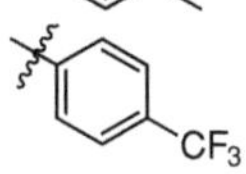	56[c]
4	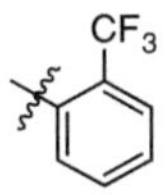	85[c]
5	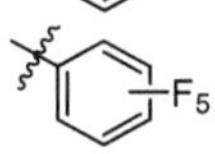	100 (91)
6	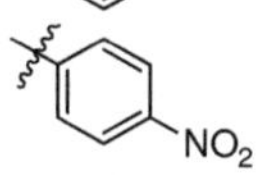	17
7	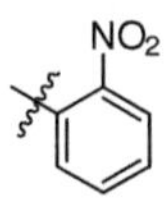	6

[a] Aryl sulfonate (0.15 mmol), butyl vinyl ether (0.60 mmol), Cy_2NMe (0.45 mmol), DPPF (5 mol%), $Ni(COD)_2$ (5 mol%) dioxane 1 mL at 100 °C for 20 h. [b] Based on ^{1}H NMR analysis. Isolated yields in parenthesis. [c] Mixtures of regioisomers were observed.

regioisomers were obtained, suggesting that both the neutral and the cationic pathway are operating. Introducing a methyl group in the *para*-position impeded the reaction, as illustrated by the reaction of 2-naphthyl tosylate (entry 2). In contrast, electron withdrawing substituents such as trifluoromethyl groups favored the nickel catalyzed Heck coupling to a greater extent although mixtures of regioisomers were once again observed (entries 3 and 4). On the other hand, employing pentafluorophenyl sulfonate as the leaving group resulted in a clean reaction, generating the α-substituted *n*-butyl vinyl ether selectively, and a 91 % yield of the corresponding 2-acetonaphthone could be secured (entry 5). Reduced catalytic activity was observed, when nitro groups were introduced on the phenylsulfonate (entries 6 and 7). Furthermore, halide additives such as LiCl and TBAB completely inhibited the reaction presumably due to trapping of the cationic nickel(II) complex, hence occupying the vacant site on the nickel nucleus required for the olefin insertion and subsequent β-hydride elimination (results not show) [37].

Table 6.4 Nickel catalyzed heck reaction various olefins

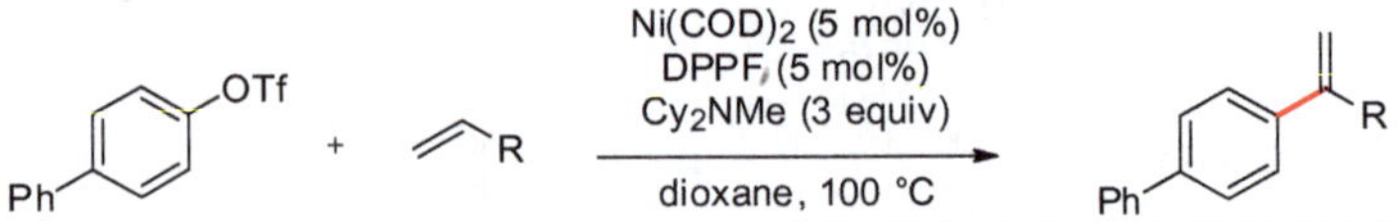

Entry[a]	Olefin	Conversion (%)[b]	Entry[a]	Olefin	Conversion (%)[b]
1	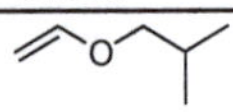	100 (91)	5	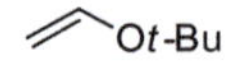	10
2		37	6	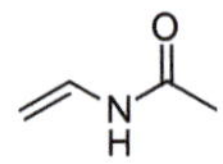	<5
3	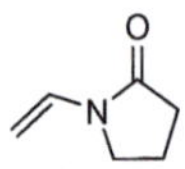	35	7	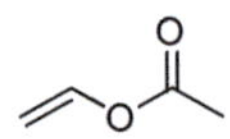	<5
4	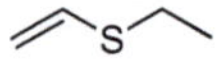	11	8	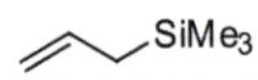	<5

[a] Aryl triflate (0.15 mmol), olefin (0.60 mmol), Cy_2NMe (0.45 mmol), DPPF (5 mol%), $Ni(COD)_2$ (5 mol%) dioxane 1 mL at 100 °C for 20 h. [b] Isolated yields

Reprinted (adapted) with permission from *J. Am. Chem. Soc.* **2012**, *134*, 443. Copyright 2011 American Chemical Society

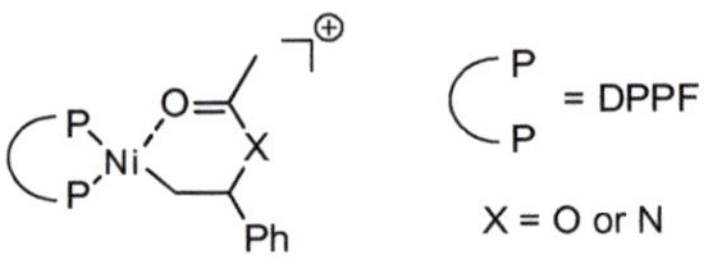

Scheme 6.3 Coordination of carbonyls to the cationic alkylnickel(II) intermediate. Reprinted (adapted) with permission from *J. Am. Chem. Soc.* **2012**, *134*, 443. Copyright 2011 American Chemical Society

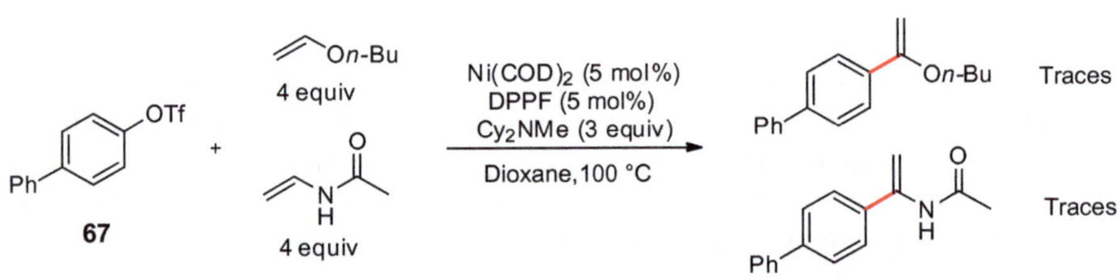

Scheme 6.4 Competition experiment between butyl vinyl ether and *N*-vinyl acetamide. Reprinted (adapted) with permission from *J. Am. Chem. Soc.* **2012**, *134*, 443. Copyright 2011 American Chemical Society

Next, different olefins were tested in the coupling of 4-biphenyl triflate (Table 6.4). Introducing an isobutyl side chain on the vinyl ether did not affect the catalytic efficiency forming the 4-phenyl acetophenone in excellent yields after hydrolysis (entry 1). Employing a sterical encumbered vinyl ether or the cyclic olefin 2,3-dihydrofuran lowered the yield dramatically (entries 2 and 3). Moreover, essentially no reaction was obtained in the coupling of *N*-vinyl acetamide, *N*-vinyl 2-pyrrolidone and vinyl acetate (entries 4–6). Possibly, a coordination of the amide/ester carbonyl group to the empty site on the metal center could potentially account for the lack of reactivity [38]. The six-membered metalocyclic cationic alkylnickel(II) complex formed after the olefin insertion, being electron deficient and unsaturated, would thereby be stabilized and hence reluctant to undergo β-hydride elimination (Scheme 6.3). Finally, the couplings of ethyl vinyl sulfide and allyltrimethylsilane failed most likely due to catalyst poisoning by the sulfide and lack of electron density in the π-system, displayed by the less electron rich allyltrimethylsilane (entries 7 and 8) [39]. These results clearly demonstrate vinyl ethers to be the most suitable olefins in the nickel catalyzed Heck reaction, contrasting the results obtained in the palladium catalyzed Heck reactions of 2-pyridyl tosylates, demonstrating the *N*-vinyl acetamides to be considerably more reactive (see Sect. 5.3). Potentially, the high selectivity could be exploited in the orthogonal functionalization of systems holding multiple olefinic moieties possessing different electronic properties. Hence, a competition experiment was set up adding a 1:1 mixture of *n*-butyl vinyl ether and *N*-vinyl acetamide to a mixture of 4-biphenyl triflate and the DPPF ligated nickel catalyst in combination with Cy_2NMe (Scheme 6.4). Only traces of the two possible 1,1-disubstituted alkenes were detected indicating an obstruction of the nickel catalyst by the *N*-vinyl acetamide demonstrating better coordinating abilities.

The results presented in Table 6.2, 6.3 and 6.4 suggest an unsaturated cationic nickel(II) intermediate to be essential in order to achieve successful coupling.

The reductive elimination in transition metal catalysis has been shown to be influenced by a number of factors including the electronic properties of the metal center and the steric effects of the coordinating ligands [40, 41]. For instance, electron poor transition metals tend to be more prone to facilitate the reductive elimination in contrast to metals with high electron densities [41–43]. Furthermore, Brown and coworkers have demonstrated, that the bite angle of bidentate ligands has a great impact on the catalyst regenerating step increasing the rate with increasing bite angles [44]. Thus, the electron deficient nature of the cationic hydridonickel intermediate and the actual bite angle of the bidentate ligand could possibly account for a more favorable β-hydride elimination and reductive elimination, thereby representing a new strategy in the nickel catalyzed Heck reaction.

In order to examine the influence of the intermediary cationic nickel(II) complexes on the elementary reactions constituting the catalytic cycle, a DFT study was performed [45]. Ni(DPPP), resembling the catalytically active DPPF derived nickel(0) catalyst, was chosen as the starting model complex in combination with phenyl triflate and ethyl vinyl ether. The bisphosphine nickel(0) complex coordinated to the phenyl triflate prior to the oxidative addition (complex **I**) was selected as set point. In the following, the free energies are given for the total system relative to this complex. (see Computational Details for further information).

The oxidative addition of phenyl triflate to the nickel(0) complex resulting in the bisphosphine phenylnickel(II) triflate complex **II** was found to be highly

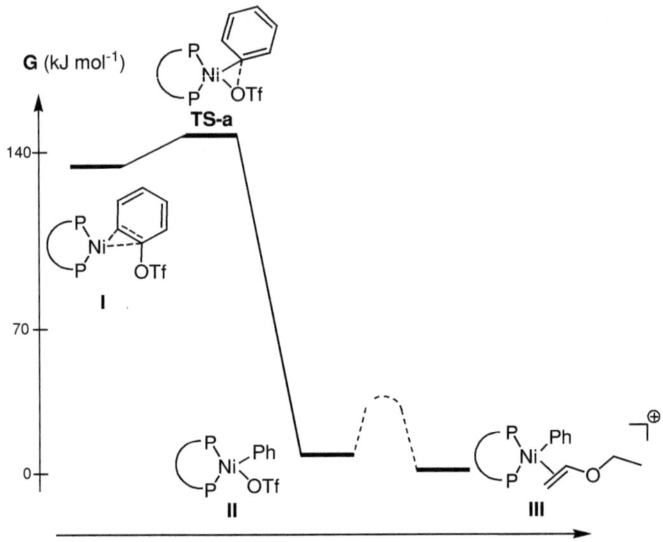

Fig. 6.1 Oxidative addition and ligand exchange. Reprinted (adapted) with permission from *J. Am. Chem. Soc.* **2012**, *134*, 443. Copyright 2011 American Chemical Society

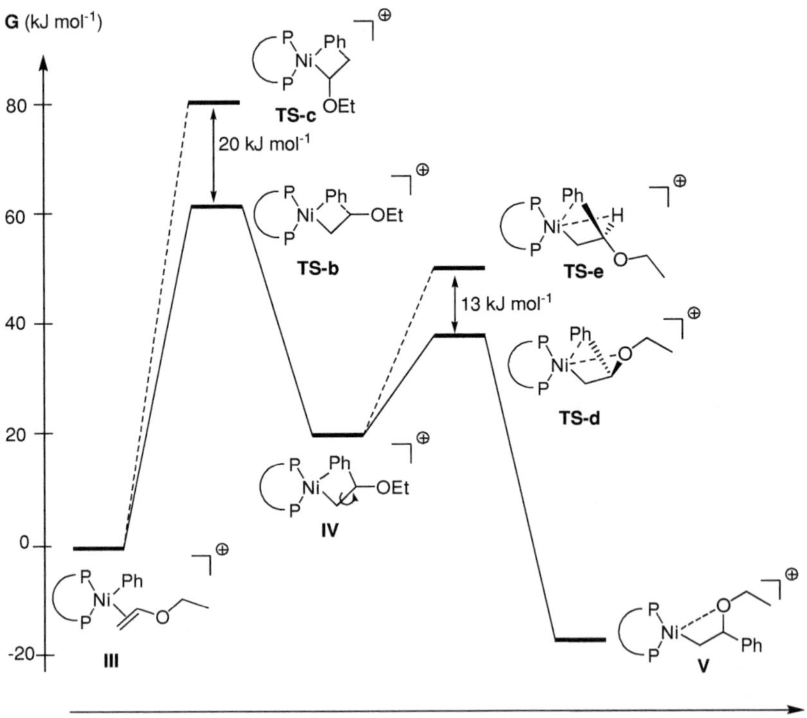

Fig. 6.2 Carbonickelation and chelation. Reprinted (adapted) with permission from *J. Am. Chem. Soc.* **2012**, *134*, 443. Copyright 2011 American Chemical Society

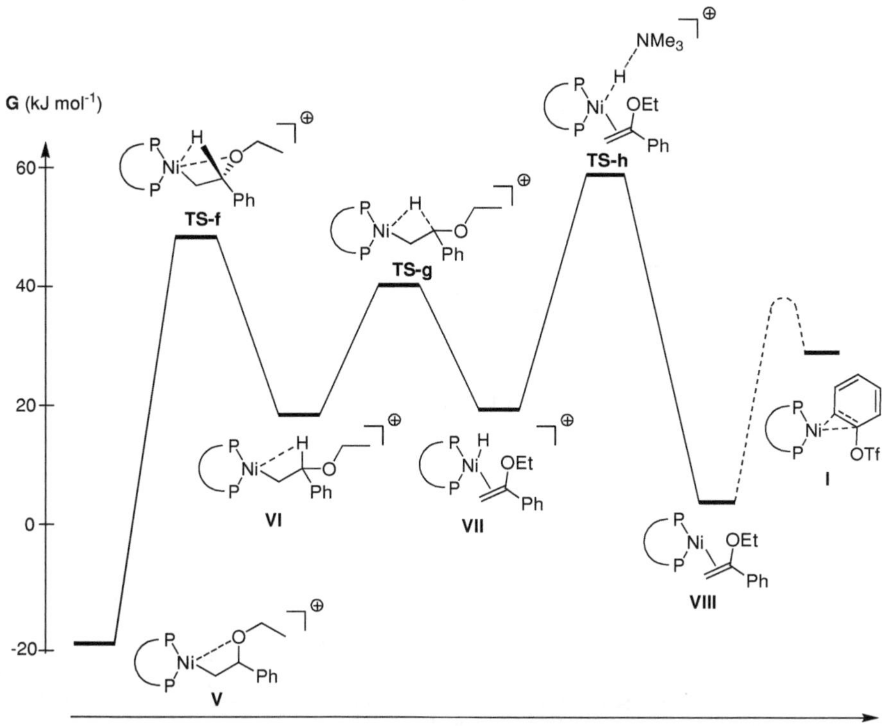

Fig. 6.3 *β*-hydride elimination and reductive elimination. Reprinted (adapted) with permission from *J. Am. Chem. Soc.* **2012**, *134*, 443. Copyright 2011 American Chemical Society

exergonic (–131 kJ mol^{-1}) with an activation barrier of 15 kJ mol^{-1} (Fig. 6.1). The exchange of the weakly coordinating triflate anion forming the cationic nickel(II) complex **III** coordinated by the ethyl vinyl ether revealed itself slightly more stable than the parent neutral nickel(II) complex **II**. As described in Sects. 1.4.2 and 1.7.1 the regioselective outcome of the reaction is determined in the subsequent carbonickelation of the olefin forming the *α*- or *β*-substituted alkene via **TS-b** and **TS-c**, respectively (Fig. 6.2). However, under the applied reaction conditions the primary alkylnickel(II) complex **TS-b** leading to the branched *α*-arylated alkene was strongly favored over the secondary alkylnickel(II) species **TS-c** by 20 kJ mol^{-1} resulting in less than 0.1 % of the linear product, in agreement with the results observed experimentally. Displacement of the coordinated phenyl group in complex **IV** with the oxygen moiety, via a rotation of the C–C bond, was found to proceed with a barrier of 18 kJ mol^{-1} compared to 31 kJ mol^{-1} leading to the agostic complex through **TS-e**. The generation of the kinetically favored oxygen coordinated cationic alkylnickel(II) complex **V** represents the most stable complex obtained after olefin insertion with a free energy of −37 kJ mol^{-1} relative to complex **IV**. The formation of the agostic nickel(II) complex **VI** by a consecutive rotation of the C–C bond in **V** proved to be a

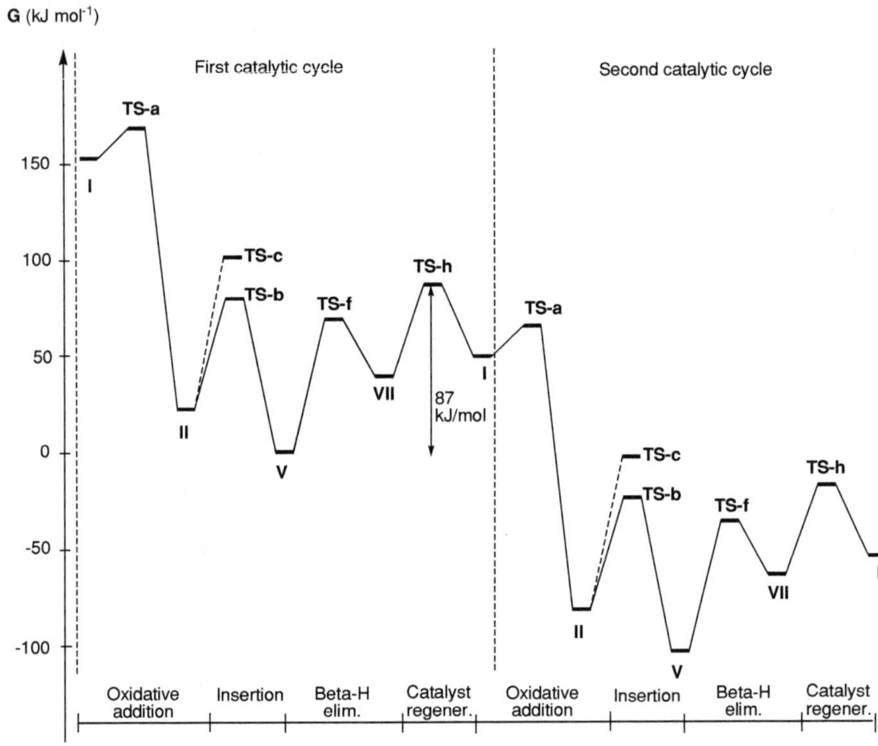

Fig. 6.4 Simplified free energy surface of two sequential catalytic cycles. Reprinted (adapted) with permission from *J. Am. Chem. Soc.* **2012**, *134*, 443. Copyright 2011 American Chemical Society

demanding process, displaying a barrier of 68 kJ mol^{-1} (Fig. 6.3). Interestingly, this C–C bond rotation **TS-f** demonstrated to be of higher energy than the following β-hydride elimination via **TS-g**, in accordance to the observations reported by Norrby and coworkers in the palladium catalyzed Heck reaction [46].

The cationic nickel(II) hydride complex **VII** then undergoes a base induced reductive elimination through **TS-h**, regenerating the nickel(0) catalyst coordinated by the α-arylated vinyl ether product **VIII**. The barrier for this process was calculated to be 47 kJ mol^{-1} using the computationally more simplistic trimethyl amine as model base compared to Cy_2NMe. Several different base promoted hydrogen abstractions were investigated including the reductive elimination from an amine coordinated hydridonickel(II) complex [47]. However, none of these were found to be more efficient, than the bimolecular process depicted as **TS-h** in Fig. 6.3. Endergonic ligand exchange (30 kJ mol^{-1}) substituting the olefinic product with another phenyl triflate completes the catalytic cycle.

From the simplified overall catalytic cycle depicted in Fig. 6.4, two transitions states can be identified as effectively irreversible (**TS-a** and **TS-h**), as they are

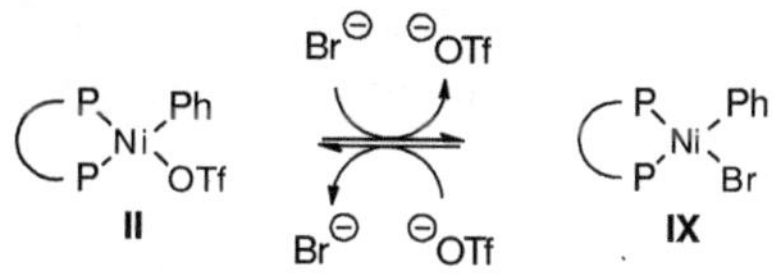

Scheme 6.5 Nickel(II) bromide and triflate complexes. Reprinted (adapted) with permission from *J. Am. Chem. Soc.* **2012**, *134*, 443. Copyright 2011 American Chemical Society

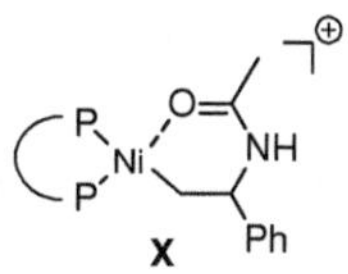

Scheme 6.6 Coordination of carbonyl oxygen. Reprinted (adapted) with permission from *J. Am. Chem. Soc.* **2012**, *134*, 443. Copyright 2011 American Chemical Society

higher than any subsequent point on the global free energy surface. The resting state of the nickel catalyzed Heck reaction can be identified as the alkylnickel(II) complex **V** generated after the migratory insertion, which would imply, that the migratory insertion is a reversible step. However, the alternative reaction pathway forming a terminal alkylnickel(II) complex is energetically inaccessible since **TS-c** is higher in energy than **TS-h** [48]. On the other hand, the β-hydride elimination and catalyst regeneration were both found to be endergonic and thus constitute the most demanding step with an overall barrier of 87 kJ mol^{-1} through **TS-h**. These findings parallel the calculations performed by Guo and coworkers also suggesting the reductive elimination to be the rate-determining step in the nickel catalyzed Heck reaction [28]. The highly exergonic oxidative addition ensures the total lowering in the energy facilitating the consecutive catalytic cycle.

On the other hand, impeded reaction rates could possibly be encountered, if any of the other potential resting states, including complex **II** obtained after the oxidative addition, were significantly stabilized (Fig. 6.1). During the investigations on the influence of the leaving group a detrimental effect of added halides to the reaction mixture was observed [49]. This effect was proposed to originate from the trapping of the crucial cationic nickel(II) intermediate blocking the vacant site mandatory in order to facilitate the olefin coordination. Consequently, calculations regarding the dissociation of a bromide anion from the nickel(II) center were conducted (Scheme 6.5). The presence of bromide anions proved to stabilize the neutral nickel(II) complex **IX** considerably by a free energy of 69 kJ mol^{-1} compared to complex **II**, which would result in a carbonickelation barrier corresponding to 132 kJ mol^{-1} essentially preventing further reaction.

Performing the nickel ctalayzed Heck reaction with olefins containing carbonyl groups such as *N*-vinylacetamide afforded the desired product only to a limited extent (Table 6.4). The potential coordination of the carbonyl oxygen to the electron deficient cationic nickel(II) intermediate was suggested to obstruct the β-hydride elimination. Hence, DFT calculations were commissioned in order to evaluate the coordinating abilities of the carbonyl moiety (Scheme 6.6).

The transition state for the β-hydride elimination from the cyclic cationic alkylnickel(II) complex **X** was located with a free energy barrier of 94 kJ mol^{-1}

compared to 68 kJ mol^{-1} (from **V** to **VII**, Fig. 6.4), accounting for the poor catalytic outcome observed using carbonyl containing olefins.

6.4 Conclusion and Outlook

A new efficient protocol for the nickel catalyzed Heck reaction of aryl triflates with vinyl ethers has been presented. Mild reaction conditions, that equal those of the corresponding palladium catalyzed Heck reaction, were applied representing a practical and more sustainable alternative to the conventional regioselective arylation of vinyl ethers. A catalytic system comprised of $Ni(COD)_2$ and DPPF in combination with the tertiary amine Cy_2NMe turned out effective in the regioselective olefination of a wide range of aryl triflates. Both electron deficient and electron rich arenes proved compatible,and the corresponding aryl methyl ketone could be secured after hydrolysis in yields approaching quantitative. Good functional group tolerance was observed matching the characteristics of the analogous palladium catalyzed Heck reaction. The high levels of catalytic activity were attributed to the intermediacy of a cationic nickel(II) complex potentially responsible for the successive β-hydride elimination and base promoted catalyst regeneration. Although these elementary reactions are normally considered challenging, DFT calculations suggested this cationic pathway to be viable in the nickel catalyzed Heck reaction.

Further development could possibly include electrophiles based on carbamates or carbonates known to readily dissociate from the nickel nucleus as demonstrated by Jamison and coworkers [29]. Additionally, introduction of Lewis acids could perhaps allow for the implementation of carbonyl containing olefins circumventing the deleterious coordinating effects of such olefins.

References

1. Affo W, Ohmiya H, Fujioka T, Ikeda Y, Nakamura T, Yorimitsu H, Oshima K, Imamura Y, Mizuta T, Miyoshi K (2006) J Am Chem Soc 128:8068
2. Lohre C, Dröge T, Wang C, Glorius F (2011) Chem Eur J 17:6052
3. Jana R, Pathak TP, Sigman MS (2011) Chem Rev 111:1417
4. Takachi M, Kita Y, Tobisu M, Fukumoto Y, Chatani N (2010) Angew Chem Int Ed 49:8717
5. Montgomery J (2004) Angew Chem Int Ed 43:3890
6. Saya L, Bhargava G, Navarro MA, Gulías M, Lopéz F, Fernández I, Castedo L, Mascareñas JL (2010) Angew Chem Int Ed 49:9886
7. Koyama I, Kurahashi T, Matsubara S (2009) J Am Chem Soc 131:1350
8. Yang C–M, Jeganmohan M, Parthasarathy K, Cheng C–H (2010) Org Lett 12:3610
9. Namitharan K, Pitchumani K (2010) Eur J Org Chem 411
10. Auvinet A–L, Harrity JPA (2011) Angew Chem Int Ed 50:2769
11. Morimoto M, Miura T, Murakami M (2010) J Am Chem Soc 132:54
12. Molinaro C, Montgomery TF (2003) J Am Chem Soc 125:8076
13. Li W, Chen N, Montgomery J (2010) Angew Chem Int Ed 49:8712

14. Everson DA, Shrestha R, Weix DJ (2010) J Am Chem Soc 132:920
15. Sergeev AG, Hartwig JF (2011) Science 332:439
16. Guan B–T, Xiang S–K, Wang B–Q, Sun Z–P, Wang Y, Zhao K–Q, Shi Z–J (2008) J Am Chem Soc 130:3268
17. Yu D–G, Li B–J, Zheng S–F, Guan B–T, Wang B–Q, Shi Z–J (2010) Angew Chem Int Ed 49:4566
18. Shimasaki T, Tobisu M, Chatani N (2010) Angew Chem Int Ed 49:2929
19. Quasdorf KW, Antoft-Finch A, Liu P, Silberstein AL, Komaromi A, Blackburn T, Ramgren SD, Houk KN, Snieckus V, Garg NK (2011) J Am Chem Soc 133:6352
20. Rosen BM, Quasdorf KW, Wilson DA, Zhang N, Resmerita A-M, Garg NK, Percec V (2011) Chem Rev 111:1346
21. Li B–J, Yu D–G, Sun C–L, Shi Z–J (2011) Chem Eur J 17:1728
22. Frisch AC, Beller M (2005) Angew Chem Int Ed 44:674
23. Glorius F (2008) Angew Chem Int Ed 47:8347
24. Weissermel K, Arpe H–J (1997) Industrial organic chemistry, 3rd edn. Wiley, Weinheim
25. Ng S–S, Ho C–Y, Jamison TF (2006) J Am Chem Soc 128:11513
26. Inamoto K, Kuroda J, Hiroya K, Noda Y, Watanabe M, Sakamoto T (2006) Organometallics 25:3095
27. Inamoto K, Kuroda J, Danjo T, Sakamoto T (2005) Synlett 1624
28. Lin B–L, Liu L, Fu Y, Luo S–W, Chen Q, Guo Q–X (2004) Organometallics 23:2114
29. Matsubara R, Jamison TF (2010) J Am Chem Soc 132:6880
30. Hansen AL, Skrydstrup T (2005) Org Lett 7:5585
31. Birkholz (née Gensow) M–N, Freixa Z, van Leeuwen PWNM (2009) Chem Soc Rev 38:1099
32. Kwong FK, Lai CW, Chan KS (2001) J Am Chem Soc 123:8864
33. Cho SH, Yoon J, Chang S (2011) J Am Chem Soc 133:5996
34. Bertrand F, Basketter DA, Roberts DW, Lepoittevin J-P (1997) Chem Res Toxicol 10:335
35. Arango V, Robledo S, Séon-Méniel B, Figadére B, Cardona W, Sáez J, Otálvaro F (2010) J Nat Prod 73:1012
36. Palladium catalyzed double bond isomerization has been reported see ref [50]
37. It has been shown that ionic liquids in combination with high temperatures and prolonged reaction times are mandatory in the nickel catalyzed Heck coupling of aryl bromides and butyl vinyl ether. See Sun L, Pei W, Shen C (2006) J Chem Res 388. For palladium catalyzed Heck reaction of aryl bromides in ionic liquids with added halides. See ref [51]
38. Curley JJ, Kitiachvili KD, Waterman R, Hillhouse GL (2009) Organometallics 28:2568
39. Olofsson K, Larhed M, Hallberg A (1998) J Org Chem 63:5076
40. Hartwig JF (2010) Organotransition metal chemistry: from bonding to catalysis. University Science Books, Sausalito
41. Hartwig JF (2007) Inorg Chem 46:1936
42. Lanci MP, Remy MS, Kaminsky W, Mayer JM, Sanford MS (2009) J Am Chem Soc 131:15618
43. Furuya T, Benitez D, Tkatchouk E, Strom AE, Tang P, Goddard WA III, Ritter T (2010) J Am Chem Soc 132:3793
44. Brown JM, Guiry PJ (1994) Inorg Chim Acta 220:249
45. The DFT calculations were performed by the group of Professor Per-Ola Norrby
46. Ambrogio I, Fabrizi G, Cacchi S, Henriksen ST, Fristrup P, Tanner D, Norrby P-O (2008) Organometallics 27:3187
47. Hills ID, Fu GC (2004) J Am Chem Soc 126:13178
48. Other ways of analyzing free energy surfaces has been presented. See Shaik KS (2011) Acc Chem Res 44:101
49. Ruan J, Xiao J (2011) Acc Chem Res 44:614
50. Gauthier D, Lindhardt AT, Olsen EPK, Overgaard J, Skrydstrup T (2010) J Am Chem Soc 132:7998
51. Ruan J, Xiao J (2011) Acc Chem Res 44:614

[illegible]
[illegible]
[illegible] J Am Chem Soc [illegible]
[illegible]
[illegible] Chem Commun [illegible]
[illegible] J Am Chem Soc [illegible]
[illegible]
[illegible] Chem Rev [illegible]
[illegible]
[illegible] Kang J, Kim J (2012) J Am Chem Soc [illegible]
[illegible] Rev [illegible]

Chapter 7
Carbonylative Heck Reaction

7.1 Introduction

Carbonylative transition metal catalysis has since the pioneering work performed by Heck and coworkers in the early 1970s evolved into a highly attractive strategy for the incorporation of the carbonyl unit using mild reaction conditions [1, 2]. The introduction of the highly versatile CO moiety not only elongates the expanding molecule with a one carbon fragment, but may also serve as a flexible platform for further manipulations of the molecule displaying numerous applications. Otherwise, the carbonyl entity may also act as a valuable motif in itself and is found in a vast number of bioactive compounds.

Palladium based carbonylation reactions employing high pressures attaining 50 bar of CO, ensured by the use of autoclaves, are not uncommon [2, 3]. Although synthetically more attractive protocols have been developed by the groups of Beller, Buchwald and others, the handling of carbon monoxide, notoriously known as a toxic, flammable and teratogenic gas, requires certain safety precautions, which consequently limits its applications [4–6]. Alternatively, solid carbon monoxide precursors such as transition metal carbonyl complexes including $Mo(CO)_6$, $W(CO)_6$ and $Fe(CO)_5$ generating CO in situ have been studied to a great extent by the group of Larhed among others [7, 8]. Stoichiometric amounts of these metal complexes are typically required, thereby lowering the gaseous quantities compared to the carbonylation reactions relying on carbon monoxide from pressurized tanks. However, the presence of metal impurities may lead to severe purification issues and interfere with the parallel CO consuming reaction [9].

Ideally, carbonylative protocols based on non-transition metal derived CO precursors, producing only the required amounts of the hazardous gas needed, would be of high interest. Furthermore, a setup easily handled, allowing an efficient purification of the desired compound, would provide the research chemist a convenient solution for the implementation of carbon monoxide as a powerful and simple installed reagent.

T. M. Gøgsig, *New Discoveries on the β-Hydride Elimination*, Springer Theses,
DOI: 10.1007/978-3-642-32099-6_7,

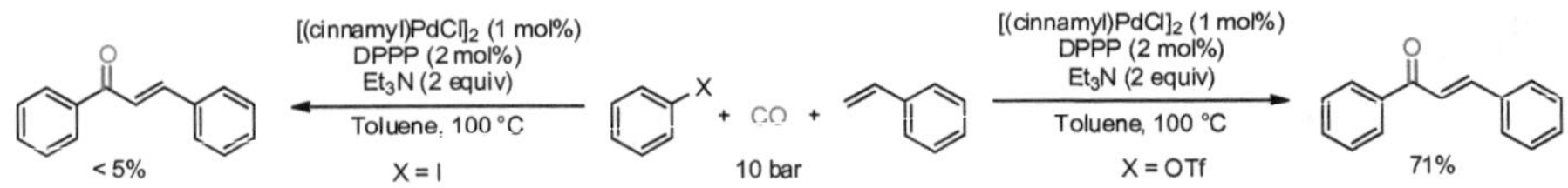

Scheme 7.1 Carbonylative Heck reactions

7.2 Background

As described in the previous chapters, new variations and improvements of the Nobel Prize decorated Heck reaction are published on a regular basis. In addition, reports revealing the mechanistic details of this highly applicable palladium catalyzed reaction have emerged representing invaluable guidelines for further developments [10–12].

Since its discovery in the mid 1990s by Miura and coworkers the carbonylative Heck reaction has only been addressed sporadically [13]. Recently, Beller and coworkers reported the first general protocol for the multicomponent carbonylative Heck reaction of aryl triflates and halides using both electron rich and deficient olefins [14–16]. Interestingly, the catalytic system applied in the carbonylative Heck reaction of aryl triflates turned out ineffective employing the corresponding aryl iodides (Scheme 7.1). This observation could indicate, that a cationic pathway is operating, when the low coordinating triflate leaving groups are utilized demonstrating the electronic properties of the involved palladium intermediates to influence the incorporation of carbon monoxide. Indeed, Grigg and coworkers have suggested electron deficient palladium complexes to assimilate carbon monoxide more readily than the corresponding electron rich complexes due to a declined back bonding into the coordinated CO π^* orbitals. Furthermore, sterically congested palladium complexes were proposed to be more suitable for the formation of the vital acylpalladium(II) intermediate, as the increased steric effects most likely force the initially coordinated CO and aryl group closer together [2].

A plausible mechanism was proposed for the palladium catalyzed carbonylative Heck reaction of aryl halides relying on an imidazolyl phosphine ligand (Scheme 7.2) [16]. Initially, an oxidative addition of the aryl halide to the starting palladium(0) species produces the tetracoordinated square planar palladium(II) complex. Coordination of carbon monoxide and subsequent migratory insertion generates the corresponding benzoylpalladium(II) intermediate. This process is believed to proceed by a preceding dissociation of one of the dative ligands generating a neutral palladium(II) species [17]. Alternatively, dissociation of the halide or pseudohalide forming a cationic benzoylpalladium(II) intermediate, similar to the polymerization processes reported by Nozaki and coworkers, could be operative [18]. The following carbopalladation forms an alkylpalladium(II) intermediate, which upon β-hydride elimination produces the desired chalcone derivative. Finally, base promoted reductive elimination regenerates the active catalyst. Stoichiometric experiments indicated that the alkene insertion was the rate-limiting step during the catalytic cycle [16]. Consequently, large excess of

Scheme 7.2 Simplified mechanism of the carbonylative heck reaction

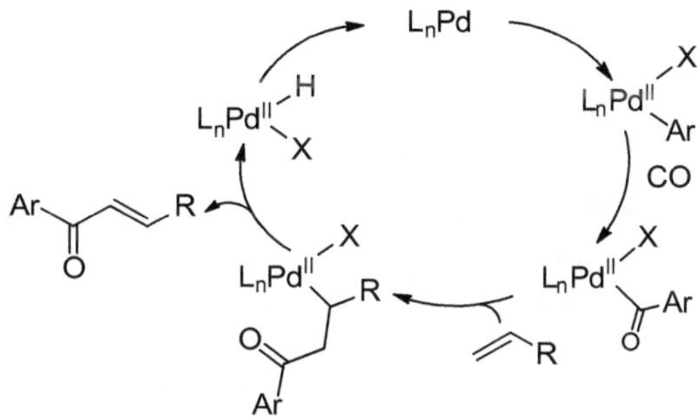

the olefin was applied in order to facilitate the carbopalladation. Furthermore, relatively high pressures of the CO gas was needed, (5–10 bar) to prevent the direct carbopalladation of the arylpalladium(II) intermediate obtained after the oxidative addition to the styrene ultimately leading to the stilbene derivatives.

Carbon monoxide possessing both σ-donating and π-accepting binding abilities may interact with the initial palladium(0) intermediate rendering it catalytically inactive especially when applied in high pressures [19, 20]. Hence, an extensive optimization is often mandatory when conducting palladium catalyzed carbonylation chemistry evaluating ligands, CO pressure, temperatures etc.

Based on the previous work performed in the Skrydstrup laboratories, regarding the aminocarbonylations utilizing carbon monoxide generated ex situ in a two-chamber system from a solid CO precursor, it was decided to investigate the carbonylative Heck reaction in further detail (Scheme 7.3) [21]. Assuming the alkene insertion being the rate determining step, as proposed by Beller and coworkers, a lowering of the CO pressure could perhaps facilitate the carbonylative Heck reaction. Furthermore, the unique setup would additionally allow for the direct installment of carbon isotopes in this class of bioactive compounds using the same strategy (Scheme 1.34, p. 27).

7.3 Results and Discussion

Based on previous work by Skrydstrup and coworkers the non-volatile and crystalline acid chloride **69**, possessing a quaternary substituent in the α-position, were chosen as the CO-precursor (Scheme 7.4). An effective delivery of the carbon monoxide could be secured upon treating **69** with a palladium catalyst based on the bulky and electron rich $P(t\text{-}Bu)_3$ in combination with DIPEA as the tertiary amine base retaining the 9-methyl-fluorene in the CO-producing chamber (Chamber A) [21, 22]. If needed, reduction of the 9-methyl-fluorene would regenerate the 9-methyl-9*H*-fluorene serving as the direct precursor for **69** and for the carbon isotope labeled analogous **69*** (Scheme 7.4). Notably, the air- and moisture stable $P(t\text{-}Bu)_3HBF_4$ salt could also be applied without any loss of reactivity, omitting glovebox techniques (Scheme 7.5).

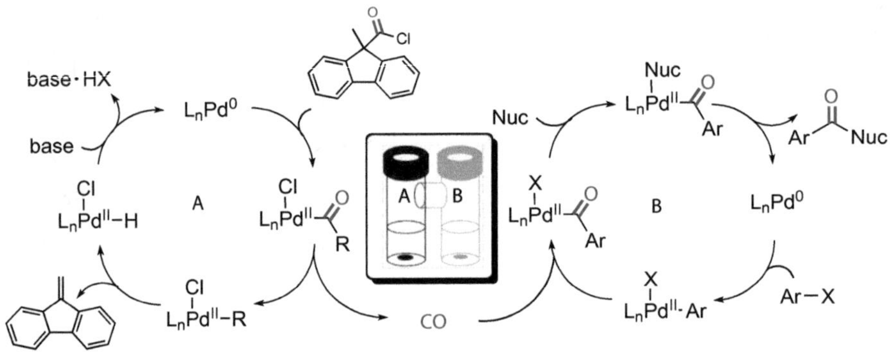

Scheme 7.3 Palladium catalyzed carbonylation chemistry using the two-chamber system

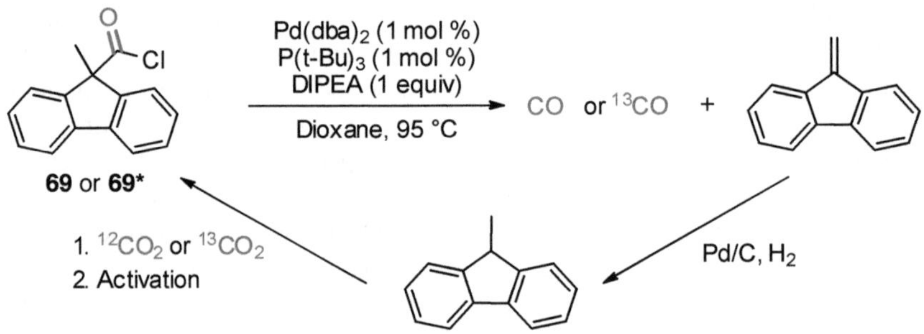

Scheme 7.4 Regeneration of CO-precursor

Initial attempts indicated the palladium chloride salt $PdCl_2$ to be suitable in the carbonylative Heck reaction of 4-iodoanisole and styrene resulting in full conversion of the starting aryl iodide after approximately 20 h in dioxane at 95 °C (Table 7.1, entry 1). However, mixtures of the desired **70** and reduced chalcone **71** adducts were detected, which could possibly be explained by the reducing properties of ether based solvents [23]. Noteworthy, the reaction proceeded without the presence of phosphine ligands using only a slight excess of carbon monoxide. A more homogeneous reaction mixture was obtained, when the more soluble cinnamylpalladium(II) chloride dimer was introduced, although similar ratios and conversion were observed (entry 4). The addition of phosphine ligands changed the catalytic outcome dramatically (entries 5–8). In particular, $P(t\text{-Bu})_3$ resulted in the direct Heck reaction, producing the stilbene derivative exclusively and preventing the incorporation of CO. However, this corresponds to the observations reported by the group of Beller in the formylation of aryl halides (entry 7) [20, 24]. Surprisingly, when a palladium catalyst relying on the geometrically comparable cataCXium A was used, the CO incorporated Heck product **70** was formed exclusively, albeit to a moderate degree (entry 8). However, decreasing the amount of added ligand formed the α-β-unsaturated ketone **70** more efficiently, indicating

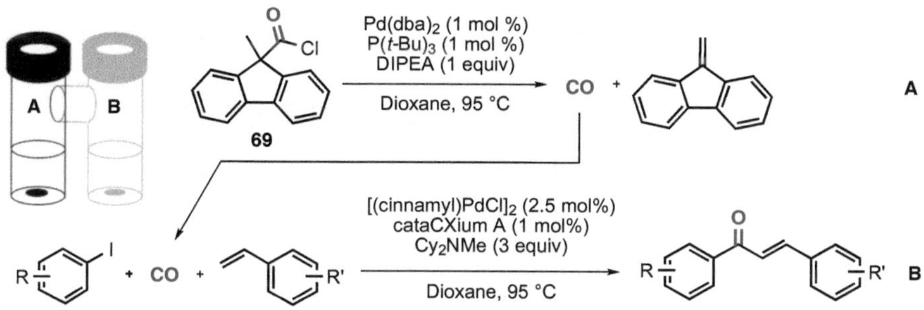

Scheme 7.5 Carbonylative Heck reaction in the two-chamber system. Reprinted (adapted) with permission from *Org. Lett.* **2011**, *13*, 2444. Copyright 2011 American Chemical Society

the catalytic system to be hampered by excess ligand (entries 8–11). Optimal reaction conditions were obtained using 1 mol% of cataCXium A resulting in excellent selectivity, forming the carbonylated Heck product **70** in 74 % isolated yield (entry 10).

With these optimized reaction conditions in hand, the generality of the catalytic system was investigated testing different aryl iodides using near stoichiometric amounts of carbon monoxide (Table 7.2). A variety of aryl iodides holding different substituents were successfully coupled to styrene forming the chalcone product in acceptable to good yields. Electron rich aryl iodides proved most adaptable in the carbonylative Heck reaction employing the two-chamber system, in agreement with the observations obtained by the groups of Heck and Miyaura (entries 1–6) [25, 26]. Introducing electron withdrawing groups and potential coordinating functionalities lowered the yield of the desired α, β-unsaturated ketone (entries 8–10). In the presence of electron withdrawing groups and *ortho*-substituents, substantial amounts of the stilbene derived products, originating from the competing direct Heck reaction, were detected (entries 8 and 11) [25, 26]. Heterocyclic compounds, represented by the tetrahydropyran and indole moieties, were well tolerated (entries 6 and 10). Although a modest yield was obtained in the coupling of 4-bromo-iodobenzene, a high degree of chemoselectivity was observed leaving the bromide unreacted (entry 7). Next, different styrene based olefins were examined in the carbonylative Heck reaction of electron rich aryl iodides using the optimized reaction conditions (Scheme 7.6). In order to address the relatively high costs of the styrene coupling partners, a prolonged reaction time could be applied, retaining a synthetically useful outcome employing only 2 equivalents of the olefin. Both electron rich and deficient styrenes coupled successfully in yields ranging from 36 to 81 %. The introduction of an *ortho*-methoxy group on the aryl iodide turned out slightly problematic, paralleling the trends observed in Table 7.2, entry 11. Nevertheless, the diuretic metochalcone **78** could be secured upon column chromatography in a satisfactory 41 % isolated yield.

Due to the flexible design of the CO-precursor and the late stage installment of the CO_2, it was decided to test the carbon isotope labeling of the pharmaceutical relevant chalcone derived compounds. Furthermore, the crystalline and easily

Table 7.1 Optimization of the carbonylative heck reaction

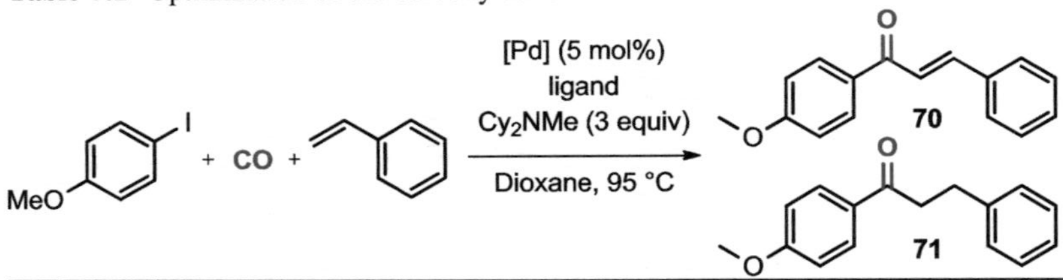

Entry[a]	[Pd]	Ligand	Ratio (**70**:**71**)	Conversion [%][b] (isolated yield) [%]
1	$PdCl_2$	–	4:1	>95
2	$Pd(OAc)_2$	–	7:2	94
3	$[(allyl)PdCl]_2$	–	4:1	>95
4	$[(cinnamyl)PdCl]_2$	–	4:1	>95 (82)
5[c]	$[(cinnamyl)PdCl]_2$	PPh_3	5:1	62
6[c]	$[(cinnamyl)PdCl]_2$	PCy_3HBF_4	3:1	38
7[c,d]	$[(cinnamyl)PdCl]_2$	$P(t\text{-}Bu)_3$	0:0	>95
8[c]	$[(cinnamyl)PdCl]_2$	cataCXium A	>95:5	51
9[e]	$[(cinnamyl)PdCl]_2$	cataCXium A	>95:5	90
10[f]	$[(cinnamyl)PdCl]_2$	cataCXium A	>95:5	>95 (74)
11[g]	$[(cinnamyl)PdCl]_2$	cataCXium A	95:5	95

[a] Chamber A: **69** (0.75 mmol), $Pd(dba)_2$ (1 mol%), $P(t\text{-}Bu)_3$ (1 mol%), DIPEA (0.75 mmol) in dioxane (3 mL) for 20 h. Chamber B: Iodoanisole (0.5 mmol), styrene (3.0 mmol), Cy_2NMe (1.5 mmol), [Pd] (5 mol%) in dioxane (3 mL) for 20 h. [b] Determined by ^1H NMR analysis. [c] Ligand (7.5 mol%). [d] Only the direct Heck coupling product was observed. [e] Ligand (2.5 mol%). [f] Ligand (1 mol%). [g] Ligand (0.5 mol%)

Reprinted (adapted) with permission from *Org. Lett.* **2011**, *13*, 2444. Copyright 2011 American Chemical Society

Table 7.2 Carbonylative heck reaction of aryl iodides in the two-chamber system

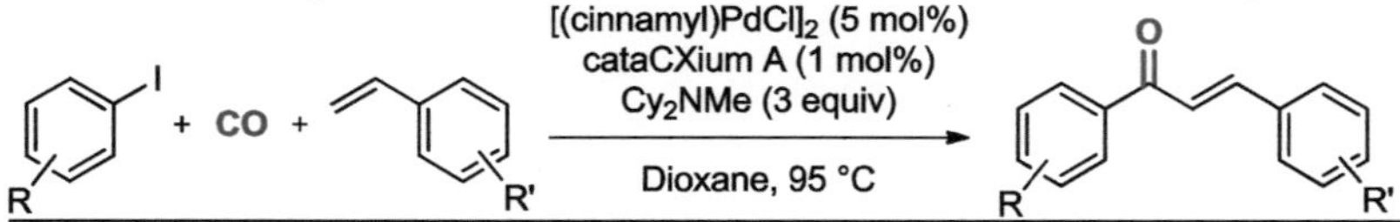

Entry[a]	Aryl iodide	Product	Yield (%)[b]
1	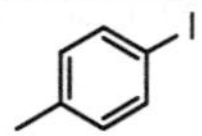	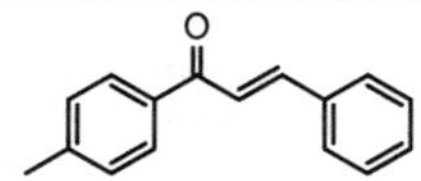	59
2	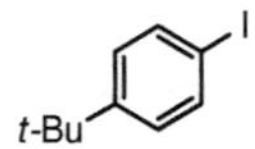	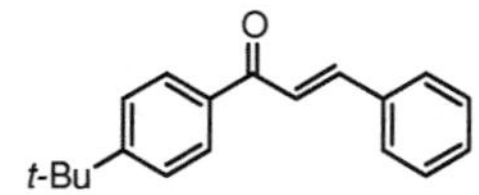	67
3	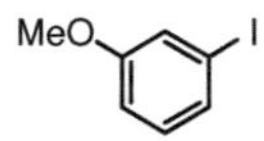	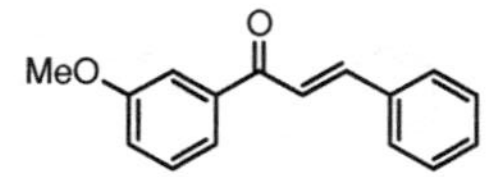	58
4	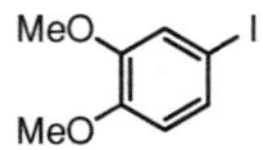	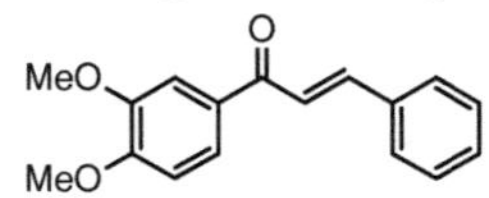	69
5	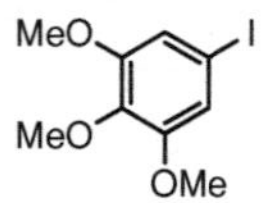	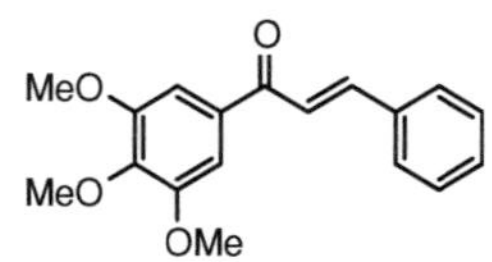	64
6	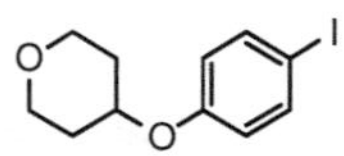	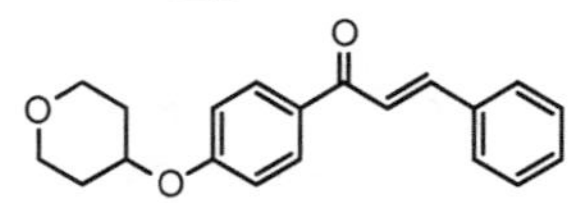	79

(continued)

Table 7.2 (continued)

Entry[a]	Aryl iodide	Product	Yield (%)[b]
7	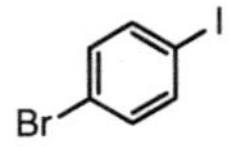	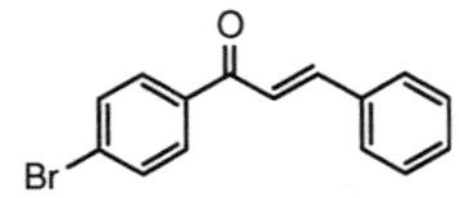	41
8	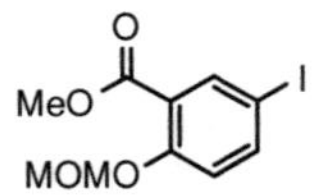	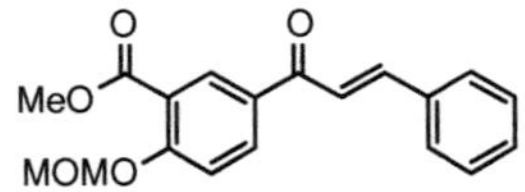	34
9	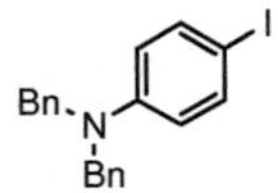	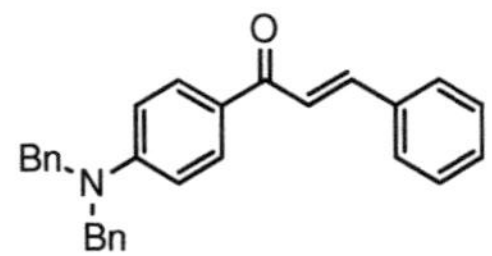	52
10	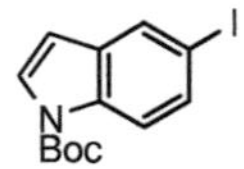	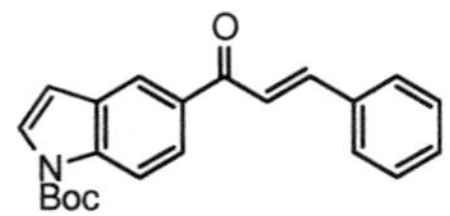	55
11	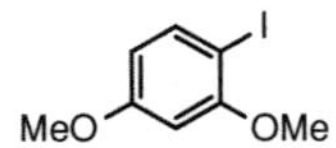	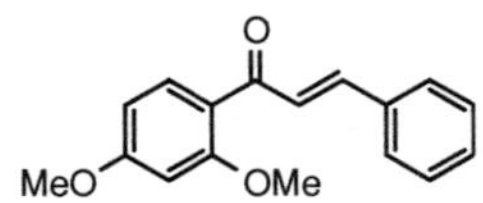	41

[a] Chamber A: **69** (0.75 mmol), $Pd(dba)_2$ (1 mol%), $P(t\text{-}Bu)_3$ (1 mol%), DIPEA (0.75 mmol) in dioxane (3 mL) for 20 h. Chamber B: Iodoanisole (0.5 mmol), styrene (3.0 mmol), Cy_2NMe (1.5 mmol), cataCXium A (1 mol%), $[(cinnamyl)PdCl]_2$ (5 mol%) in dioxane (3 mL) for 20 h. [b] Isolated yields

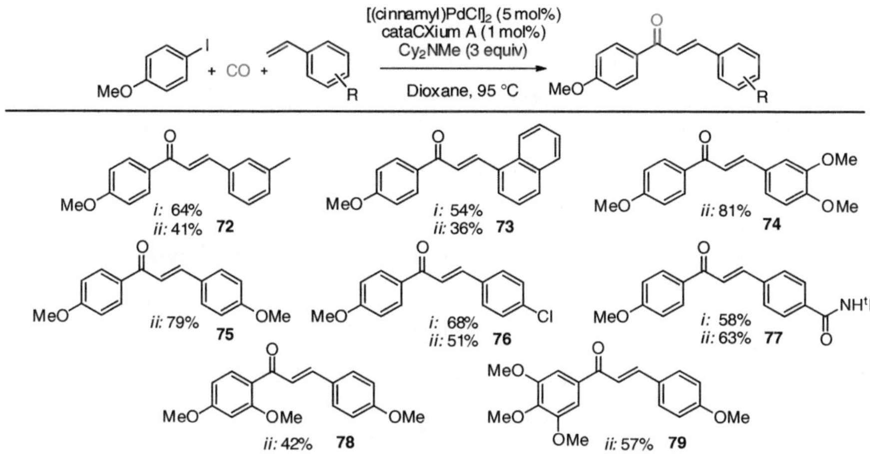

[a] Conditions *i*: Chamber A: **69** (0.75 mmol), $Pd(dba)_2$ (1 mol%), $P(t\text{-}Bu)_3$ (1 mol%), DIPEA (0.75 mmol) in dioxane (3 mL) for 20 h. Chamber B: Iodoanisole (0.5 mmol), olefin (3.0 mmol), Cy_2NMe (1.5 mmol), cataCXium A (1 mol%), $[(cinnamyl)PdCl]_2$ (5 mol%) in dioxane (3 mL) for 20 h. Conditions *ii*: Same as *i* using olefin (1 mmol) for 40 h in Chamber B.

Scheme 7.6 Carbonylative Heck Reaction of Styrene Derivatives in the Two-Chamber System. Reprinted (adapted) with permission from *Org. Lett.* **2011**, *13*, 2444. Copyright 2011 American Chemical Society

handled acid chlorides **69** and **69*** enable an accurate measured dosage of carbon monoxide, leaving this procedure attractive in the carbon isotope labeling. In this regard the ^{13}C labeled acid chloride **69***, obtained from 9-fluorenone, was employed as the limiting reagent (Scheme 7.7) [21]. Gratifyingly, the two-chamber system facilitated the incorporation of ^{13}C labeled carbon monoxide efficiently using the optimized catalytic system. Employing the highly electron rich 3,4-dimethoxystyrene in the carbonylative Heck reaction of 4-iodoanisole formed the ^{13}C labeled chalcone derivative **80** in a 73 % isolated yield with respect to the carbon monoxide precursor **69***. Furthermore, [^{13}C]metochalcone **81** was formed in a 42 % yield, comparable to the [^{12}C] analogous in Scheme 7.6, demonstrating the pressure of carbon monoxide to be of little importance. These findings imply, that the arylpalladium(II) complex formed after the oxidative addition displays a certain degree of stability. Although carbon monoxide is only partially soluble in non-polar organic solvents, it has been shown, that 1,4-dioxane accommodates the solubility of carbon monoxide to an increased extent and can be compared with polar solvents such as ethanol [27, 28]. These properties could possibly account for the successful couplings applying substoihiometric amounts of CO. Finally, 3,4,5-trimethoxyiodobenzene was transformed into an indanone structure **83** containing a ^{13}C labeled carbonyl motif in a two-steps procedure using substoichiometric amounts of ^{13}C labeled carbon monoxide.

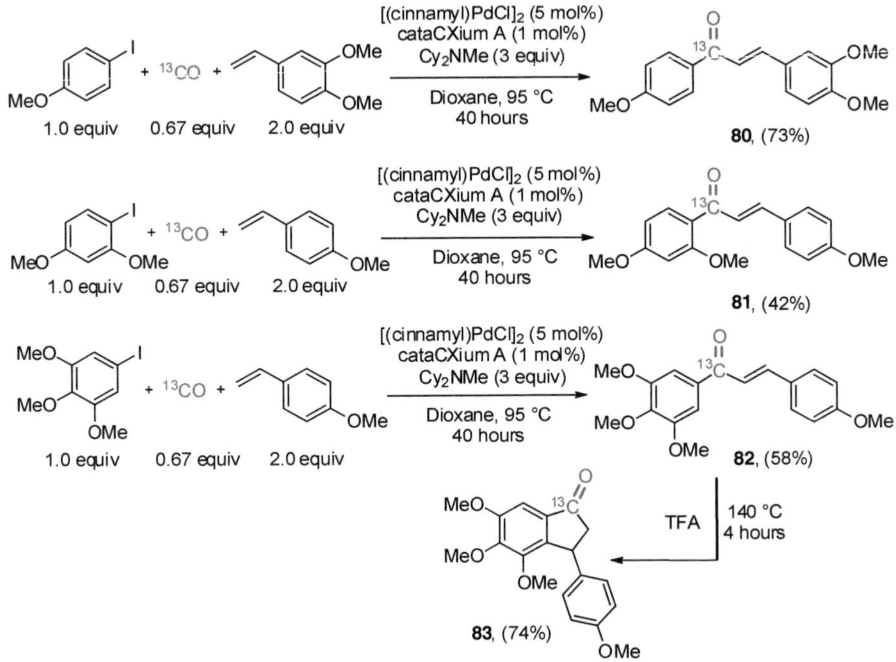

Scheme 7.7 ^{13}C Isotope labeling employing the two-chamber system. Reprinted (adapted) with permission from *Org. Lett.* **2011**, *13*, 2444. Copyright 2011 American Chemical Society

7.4 Conclusion and Outlook

In conclusion, the two-chamber system proved effective in facilitating the carbonylative Heck reaction of various aryl iodides and styrenes olefins producing the resulting chalcone derivatives in good yields.

In order to generate the desired amounts of carbon monoxide needed, a palladium based catalyst relying on the sterically encumbered and highly electron donating $P(t\text{-}Bu)_3$ ligand were employed, facilitating the central decarbonylation process efficiently. Near stoichiometric amounts of carbon monoxide gas were delivered ex situ from a stable and crystalline acid chloride. This in turn leaves the two-chamber protocol as an attractive alternative to the conventional transition metal based CO precursors and carbon monoxide canisters avoiding transition metal contaminations and high pressure equipment, respectively. In addition, ^{13}C labeling experiments were successfully accomplished forming the carbon isotope labeled chalcones in good yields applying substoichiometric amounts of [^{13}C] carbon monoxide. These results demonstrate the CO pressure to be of limited importance in the carbonylative Heck reaction when employing a catalytic system based on cataCXium A.

The development of this new two-chamber carbonylation technology enables research chemists to perform carbonylation chemistry without the often required safety precautions and specialized high-pressure equipment. Furthermore, the setup allows for the direct implementation of carbon isotope labeling, hence diverging this strategy from the conventional approaches. Especially, the ability of adding a measured amount of carbon monoxide without interfering with the carbonylative reaction poses a great advantage. Potentially, this strategy could be expanded to include the ^{14}C and ^{11}C isotope labeling of bioactive compounds utilized in metabolic pathway studies and the increasingly common PET scan studies for the diagnosis of cancer and Alzheimer's disease.

Further investigations on the adaptability of the two-chamber system in the palladium catalyzed carbonylation reactions are currently en route in the Skrydstrup laboratories and will be reported in due time.

References

1. Schoenberg A, Bartoletti I, Heck RF (1974) J Org Chem 39:3318
2. Grigg R, Mutton SP (2010) Tetrahedron 66:5515
3. Brennführer A, Neumann H, Beller M (2009) Angew Chem Int Ed 48:4114
4. Wu X-F, Jiao H, Neumann H, Beller M (2011) ChemCatChem 3:726
5. Munday RH, Martinelli JR, Buchwald SL (2008) J Am Chem Soc 130:2754
6. Arthuis M, Lecup A, Roulland E (2010) Chem Commun 46:7810
7. Kolla L (ed) (2008) Modern carbonylation methods. Wiley and Sons, Germany
8. Roberts B, Liptrot D, Alcaraz L, Luker T, Stocks M (2010) J Org Lett 12:4280
9. Wu X, Wannberg J, Larhed M (2006) Tetrahedron 62:4665
10. Amatore C, Jutand A (2000) Acc Chem Res 33:314
11. Carrow BP, Hartwig JF (2010) J Am Chem Soc 132:79
12. Henriksen ST, Norrby P-O, Kaukoranta P, Andersson PG (2008) J Am Chem Soc 130:10414
13. Satoh T, Itaya T, Okuro K, Miura M, Nomura M (1995) J Org Chem 60:7267
14. Wu X-F, Neumann H, Beller M (2010) Angew Chem Int Ed 49:5284
15. Wu X-F, Jiao H, Neumann H, Beller M (2011) ChemCatChem 3:726
16. Wu X-F, Neumann H, Spannenberg A, Schulz T, Jiao H, Beller M (2010) J Am Chem Soc 132:14596
17. Martinelli JR, Watson DA, Freckmann DMM, Barder TE, Buchwald SL (2008) J Org Chem 73:7102
18. Fujita T, Nakano K, Yamashita M, Nozaki K (2006) J Am Chem Soc 128:1968
19. Gaviño R, Pellegrini S, Castanet Y, Mortreux A, Mentré O (2001) App Catal A General 217:91
20. Sergeev AG, Spannenberg A, Beller M (2008) J Am Chem Soc 130:15549
21. Hermange P, Lindhardt AT, Taaning RH, Bjerglund K, Lupp D, Skrydstrup T (2011) J Am Chem Soc 133:6061
22. Gauthier D, Lindhardt AT, Olsen EPK, Overgaard J, Skrydstrup T (2010) J Am Chem Soc 132:7998

23. Martins A, Candito DA, Lautens M (2010) Org Lett 12:5186
24. Klaus S, Neumann H, Zapf A, Strübing D, Hübner S, Almena J, Riermeier T, Gross P, Sarich M, Krahnert W-R, Rossen K, Beller M (2006) Angew Chem Int Ed 45:154
25. Ishiyama T, Kizaki H, Hayashi T, Suzuki A, Miyaura N (1998) J Org Chem 63:4726
26. Garrou PE, Heck RF (1976) J Am Chem Soc 98:4115
27. Cao P, Sun Y, Gu R (2004) J Phys Chem B 108:4716
28. Veleckis E, Hacker DS (1984) J Chem Eng Data 29:36

Chapter 8
Conclusions

Transition metal catalysis constitutes a vital method for the construction of new C–C bonds on both a laboratory and industrial scale. In particular, palladium catalysis has established its position as one of the most powerful and reliable strategies in the syntheses of a broad spectrum of pharmaceuticals, agrochemicals, cosmetics etc. In order to develop this area of organometallic chemistry even further a full understanding of the involved mechanisms are of utmost importance.

In this dissertation an in-depth investigation of the palladium 1,2-migration have been presented. Scope and limitations of the migratory process in the Heck reaction were investigated disclosing specific requirements and trends. The 1,2-migration was expanded to include the Negishi coupling reveling related characteristics suggesting parallel mechanisms. However, stoichiometric experiments implied two distinct pathways—a neutral and a cationic—depending on the catalytic system. In both cases an unprecedented alkenyl β-hydride elimination most likely seemed to be operative, thus expanding the applications of palladium catalysis.

Furthermore, processes relying on β-hydride eliminations were explored resulting in different variations of the Heck reaction. Aryl tosylates representing attractive alternatives to the corresponding triflates were successfully applied in the regioselective Heck reaction of electron rich olefins producing highly functionalized pyridines and pyrimidines relevant to the pharmaceutical industry. A nickel catalyzed Heck reaction was developed as a competitive and more sustainable alternative to the palladium catalyzed analogous. Experimental details and DFT calculations proposed a cationic nickel intermediate to be critical for successful reaction. Finally, a palladium catalyzed carbonylative Heck reaction using carbon monoxide generated ex situ relying on a newly patented two-chamber system were accomplished. The unique setup allowed for the effective incorporation of ^{13}C labeled carbon monoxide introducing a novel strategy for the carbon isotope labeling of organic compounds.

T. M. Gøgsig, *New Discoveries on the β-Hydride Elimination*, Springer Theses, DOI: 10.1007/978-3-642-32099-6_8, © Springer-Verlag Berlin Heidelberg 2012

Chapter 8
Conclusions

Appendix: Experimental Section

Experimentals to Chap. 4

General Methods

All manipulations were conducted under an inert atmosphere in a glovebox unless otherwise stated. NMR data were collected on a Varian spectrometer. ^{1}H NMR spectra were recorded at 400 or 500 MHz and referenced relative to the applied solvent. ^{13}C NMR spectra were recorded at 100 or 125 MHz using $CDCl_3$ as an external reference. $^{31}P\{^{1}H\}$ NMR spectra were recorded at 160 or 200 MHz using H_3PO_4 as an external reference. All solvents were dried over neutral alumina in a solvent system under a positive argon pressure. *t*-Butyl vinyl tosylate **14**, [1] *t*-butyl vinyl bromide **36** [2] and *o*-tolyl vinyl-1-chloride **18** [3] were prepared from the parent ketone according to known procedures. *o*-Tolyl vinyl-2-chloride **44** were synthesized by a Hunsdiecker reaction starting from the *o*-tolyl acrylic acid [4]. Pd[P(*o*-tol)$_3$]$_2$ was prepared according to known procedures [5].

Pd(PPF-t-Bu)(t-Bu-vinyl)(Br) (Non-Migrated) 37

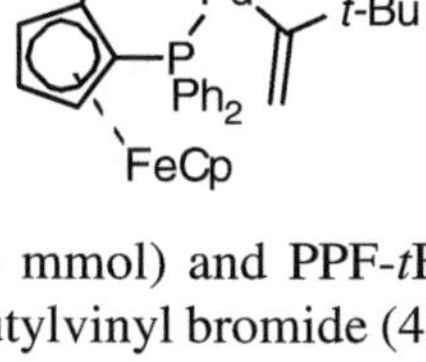

Pd[P(*o*-tol)$_3$]$_2$ (100.0 mg, 0.14 mmol) and PPF-*t*Bu (83.4 mg, 0.15 mmol) were dissolved in 5 mL THF and *t*-butylvinyl bromide (45.6 mg, 0.28 mmol) dissolved in 1 mL THF was then added to the mixture resulting in a color change from deep orange to brown orange. The reaction was complete after stirring for 10 minutes at

T. M. Gøgsig, *New Discoveries on the β-Hydride Elimination*, Springer Theses,
DOI: 10.1007/978-3-642-32099-6,

room temperature as determined by ^{31}P NMR spectroscopy. After filtration through celite, the reaction mixture was concentrated in vacuo. Addition of pentane led to the precipitation of orange solids which were collected by filtration and washed with pentane several times resulting in 81.3 mg (72% yield) of the title product as an orange solid. Crystals suitable for X-ray crystallography were obtained by pentane diffusion to a concentrated THF solution for 3 days at room temperature. This resulted in red orange crystals. ^{1}H NMR (CD_2Cl_2) δ (ppm) 8.26-8.22 (m, 2H), 7.65-7.57 (m, 5H), 7.44-7.36 (m, 3H), 4.60 (ddd, 1H, J = 2.0, 13.2, 29.6 Hz), 5.52 (bs, 1H), 4.26 (bs, 1H), 4.15 (bs, 1H), 3.61 (s, 5H), 3.44-3.36 (m, 1H), 3.21-3.35 (m, 1H), 2.05 (t, 3H, J = 7.6 Hz), 1.70 (d, 9H, J = 11.6 Hz), 1.30 (d, 9H, J = 12.8 Hz), 0.94 (s, 9H). ^{13}C NMR (CD_2Cl_2) δ (ppm) 177.4 (dd, J = 6.1, 105.5 Hz), 136.2 (d, J = 44.0 Hz), 135.7 (d, J = 13.6 Hz), 135.4 (d, J = 9.1 Hz), 131.1 (d, J = 2.3 Hz), 129.7 (d, J = 2.3 Hz), 128.5 (d, J = 11.4 Hz), 127.5 (d, J = 10.7 Hz), 116.0-118.9 (m), 96.0 (dd, J = 6.1, 18.2 Hz), 81.0 (dd, J = 9.8, 38.7 Hz), 75.6 (d, J = 3.1 Hz), 70.5, 70.0 (d, J = 7.6 Hz), 67.7 (d, J = 6.8 Hz), 43.1-43.0 (m), 39.1, 37.9, 33.3 (d. J = 4.6 Hz), 33.1 (d, J = 2.3 Hz), 32.1(d, J = 3.8 Hz), 31.4 (d, J = 4.5 Hz), 18.4 (d, J = 6.0 Hz). ^{31}P NMR (C_6D_6) δ (ppm) 60.5 (d, J = 44.0 Hz), 16.8 (d, J = 44.0 Hz). Anal calc'd for $C_{38}H_{51}BrFeP_2Pd$: C, 56.21; H, 6.33. Found: C, 56.13; H 6.44.

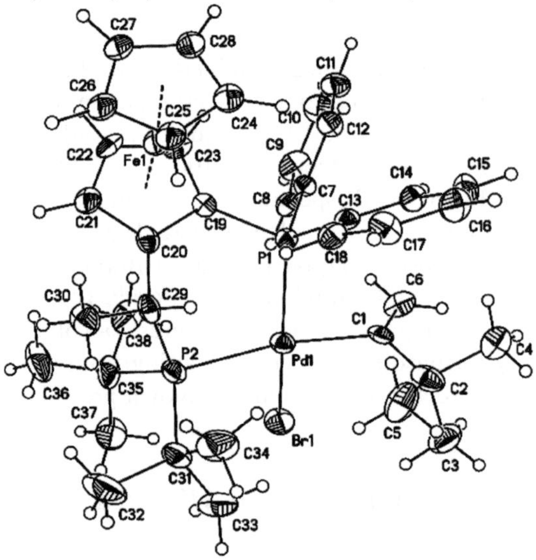

Selected intramolecular bond distances and angles

Bond distances (Å)			
Pd(1)–P(1)	2.245(2)	Pd(1)–P(2)	2.448(2)
Pd(1)–Br(1)	2.5018(10)	Pd(1)–C(1)	2.081(8)
Bond angles (deg)			
C(1)–Pd(1)–P(1)	87.6(2)	C(1)–Pd(1)–Br(1)	85.8(2)
P(1)–Pd(1)–P(2)	97.22(7)	P(2)–Pd(1)–Br(1)	96.02

Pd(PPF-t-Bu)(t-Bu-vinyl)(Br) (Migrated, Interchanged) 38

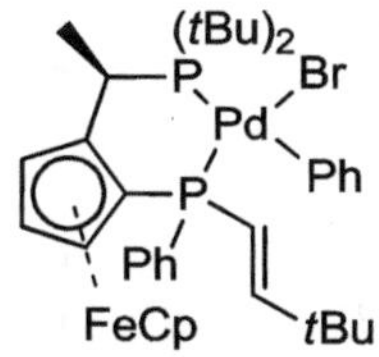

Pd[P(*o*-tol)$_3$]$_2$ (150.0 mg, 0.21 mmol) and PPF-*t*Bu (125.2 mg, 0.23 mmol) were dissolved in 5 mL toluene and *t*-butylvinyl bromide (58.7 mg, 0.23 mmol) dissolved in 1 mL toluene was then added to the mixture resulting in a color change from deep orange to brown orange. The reaction was complete after stirring for 2 h at 110 °C as determined by ^{31}P NMR spectroscopy. After filtration through celite, the reaction mixture was concentrated in vacuo. Diethyl ether was added and the product was crashed out by adding pentane and collected by filtration. After washing the crude product thoroughly with pentane, a diffusion of pentane to a concentrated toluene solution of the resulting orange solids was performed at −20 °C for 2 days. This afforded crystals suitable for X-ray crystallography as red orange crystals obtained upon filtration and subsequent washing with pentane. ^{1}H NMR (CD_2Cl_2) δ (ppm) 7.36 (dd, 1H, J = 17.0, 22.6 Hz), 7.27–7.19 (m, 5H), 6.83–6.81 (m, 2H), 6.52–6.43 (m, 3H), 6.09 (dd, 1H, J = 7.2, 17.0 Hz), 4.56 (bs, 1H), 4.27 (bs, 1H), 4.16 (s, 5H), 3.84 (bs, 1H), 3.33–3.31 (m, 1H), 2.02 (t, 3H, J = 7.6 Hz), 1.66 (d, 9H, J = 11.9 Hz), 1.32–1.29 (m, 18H). ^{31}P NMR (CD_2Cl_2) δ (ppm) 68.6 (d, J = 38.3 Hz), 12.2 (d, J = 38.3 Hz).

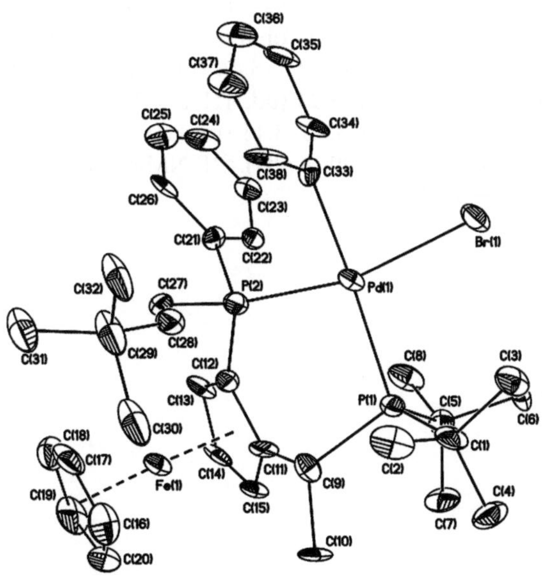

Selected Intramolecular Bond Distances and Angles

Bond distances (Å)			
Pd(1)–P(1)	2.419(3)	Pd(1)–P(2)	2.247(3)
Pd(1)–Br(1)	2.5030(16)	Pd(1)–C(33)	2.120(12)
P(2)–C(21)	1.840(12)	P(2)–C(27)	1.761(15)
Bond angles (deg)			
C(33)–Pd(1)–P(2)	84.2(3)	C(33)–Pd(1)–Br(1)	83.6(3)
P(1)–Pd(1)–P(2)	95.63(12)	P(1)–Pd(1)–Br(1)	99.06(9)

Pd(PPF-t-Bu)(t-Bu-vinyl)(OTs) (Migrated, Interchanged) 41

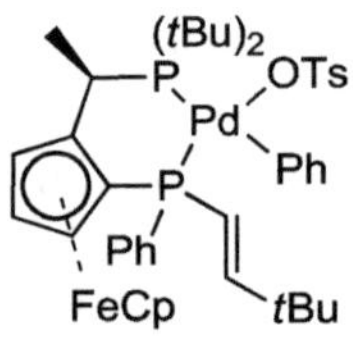

Pd[P(o-tol)$_3$]$_2$ (150.0 mg, 0.21 mmol) and PPF-t-Bu (113.8 mg, 0.21 mmol) were dissolved in 5 mL benzene and t-butylvinyl tosylate (58.7 mg, 0.23 mmol) dissolved in 1 mL benzene was then added to the mixture resulting in a color change from deep orange to brown orange. The reaction was complete after stirring for 30 minutes at 35 °C as determined by ^{31}P NMR spectroscopy. After filtration through celite, the reaction mixture was concentrated in vacuo. Diffusion of pentane to a very concentrated toluene solution of the resulting crude product was performed at −20 °C for 7 days affording crystals suitable for X-ray crystallography as red orange crystals obtained after washing several times with pentane. ^{1}H NMR (C_6D_6) δ (ppm) 7.53–8.44 (m, 2H), 7.30–7.16 (m, 4H), 7.01–6.86 (m, 4H), 6.69–6.55 (m, 4H), 5.94 (dd, 1H, J = 10.3, 17.1 Hz), 3.96 (bs, 1H), 3.79–3.76 (m, 6H), 3.59 (bs, 1H), 3.13–3.11 (m, 1H), 1.89 (s, 3H), 1.61 (d, 9H, J = 12.0 Hz), 1.49–1.46 (m, 12 H), 0.99 (s, 9H). ^{31}P NMR (C_6D_6) δ (ppm) 68.0 (d, J = 37.3 Hz), 14.2 (d, J = 37.3 Hz).

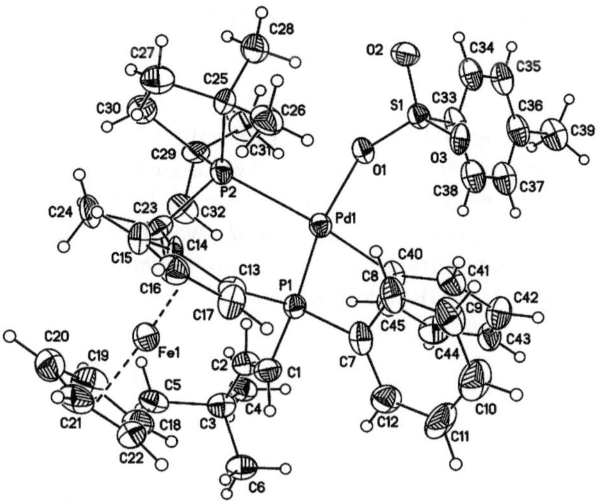

Selected intramolecular bond distances and angles

Bond distances (Å)			
Pd(1)–P(1)	2.205(3)	Pd(1)–P(2)	2.435(3)
Pd(1)–O(1)	2.152(8)	Pd(1)–C(40)	2.024(10)
P(1)–C(1)	1.777(13)	P(1)–C(7)	1.831(11)
Bond angles (deg)			
C(40)–Pd(1)–P(1)	85.1(3)	C(40)–Pd(1)–O(1)	88.0(6)
P(1)–Pd(1)–P(2)	96.71(10)	P(2)–Pd(1)–O(1)	92.4(5)

Standard Procedures for Small Scale Investigations of the 1,2-Migration

Pd[P(*o*-tol)$_3$]$_2$ (1 equiv), ligand (1 equiv) and trimesitylphosphine (1 equiv) as internal standard were dissolved in 0.5 mL solvent and stirred for 10 minutes or until all the solids were dissolved. A ^{31}P-NMR was then obtained as standard reference. The electrophile (1.3 equiv) was dissolved in 0.2 mL solvent and added to the reaction mixture by syringe. The reaction was then followed by ^{31}P-NMR and ^{1}H-NMR for several hours at the desired temperature or until complete reaction. Additives were added by syringe together with the electrophile.

Generation of the Non-Migrated Pd(PPF-t-Bu)(o-tol)Cl (Scheme 4.10)

Pd[P(*o*-tol)$_3$]$_2$ (10.0 mg, 0.014 mmol), PPF-*t*-Bu (7.6 mg, 0.014 mmol) and trimesitylphosphine (5.4 mg, 0.014 mmol) were dissolved in 0.5 mL benzene-d_6. A ^{31}P NMR was recorded. 1-(1-Chlorovinyl)-2-methylbenzene (2.8 mg, 0.018 mmol) dissolved in 0.2 mL benzene-d_6 was then added to the mixture resulting in a color change from deep orange to brown orange. The reaction was followed by ^{31}P and ^{1}H NMR analysis.

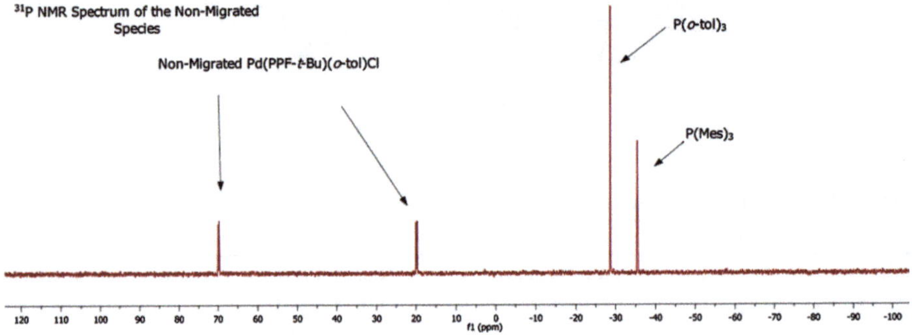

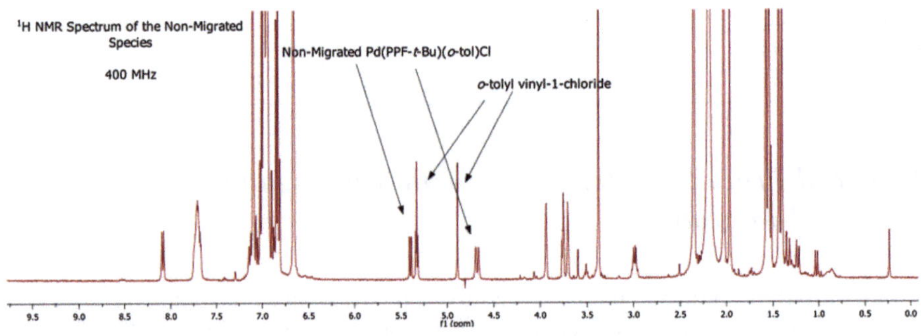

Investigation of the Effect of ZnCl$_2$ (Scheme 4.10)

Pd[P(*o*-tol)$_3$]$_2$ (10.0 mg, 0.014 mmol), PPF-*t*-Bu (7.6 mg, 0.014 mmol), trimesitylphosphine (5.4 mg, 0.014 mmol) and ZnCl$_2$ (9.5 mg, 0.070 mmol) were mixed in 0.5 mL benzene-d_6. A ^{31}P NMR was recorded. 1-(1-Chlorovinyl)-2-methylbenzene (2.8 mg, 0.018 mmol) dissolved in 0.2 mL benzene-d_6 was then added to the mixture resulting in a color change from deep orange to brown orange. The reaction was followed by ^{31}P, ^{1}H and ^{1}H{^{31}P} NMR analysis.

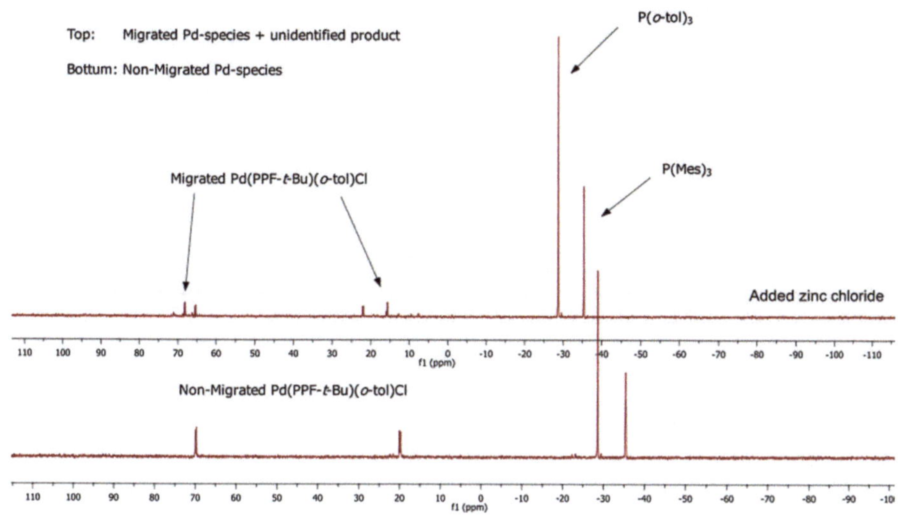

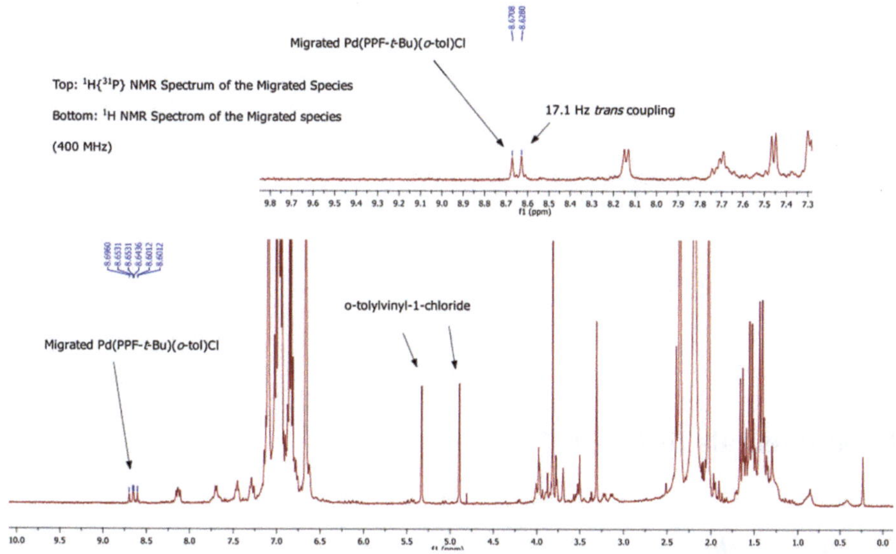

Independent Synthesis of the Migrated Pd(PPF-t-Bu)(o-tol)Cl (Scheme 4.10)

Pd[P(*o*-tol)$_3$]$_2$ (10.0 mg, 0.014 mmol), PPF-*t*-Bu (7.6 mg, 0.014 mmol), trimesitylphosphine (5.4 mg, 0.014 mmol) and $ZnCl_2$ (9.5 mg, 0.070 mmol) were mixed in 0.5 mL benzene-d_6. A ^{31}P NMR was recorded. 1-(1-Chlorovinyl)-2-methylbenzene (2.8 mg, 0.018 mmol) dissolved in 0.2 mL benzene-d_6 was then added to the mixture resulting in a color change from deep orange to brown orange. The reaction was followed by ^{31}P, ^{1}H and ^{1}H{^{31}P} NMR analysis.

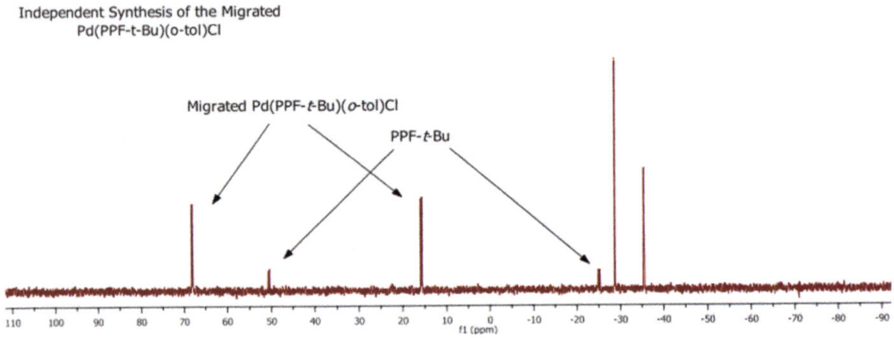

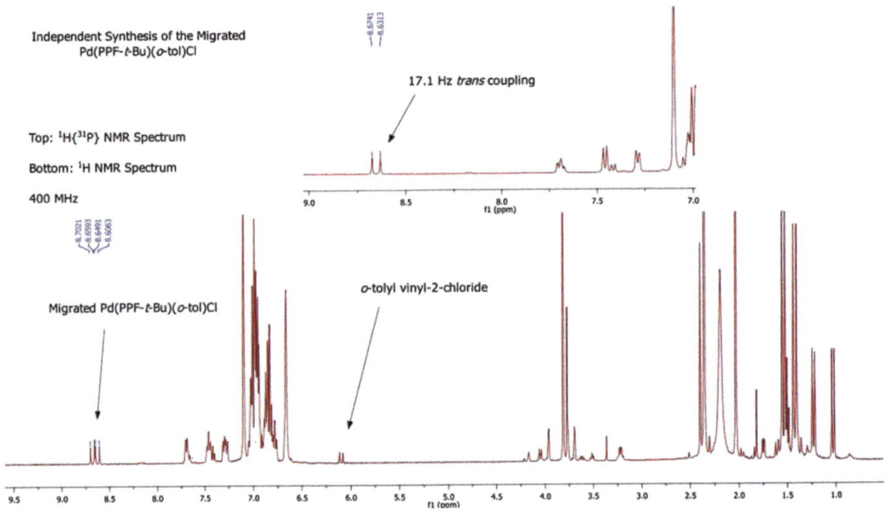

Experimentals to Chap. 6

General Methods

All reactions were carried out in 7.0 mL sample vials with a teflon sealed screwcap in a glovebox under an argon atmosphere. All purchased chemicals were used as received without further purification. Solvents were dried according to standard procedures, reactions were monitored by thin-layer chromatography (TLC) analysis and flash chromatography was carried out on silica gel 60 (230-400 mesh). The chemical shifts are reported in ppm relative to solvent residual peak [6]. The ^{1}H NMR spectra were recorded at 400 MHz, ^{13}C NMR spectra were recorded at 100 MHz and ^{19}F NMR spectra were recorded at 376 MHz on a Varian

Mercury 400 spectrometer. MS spectra were recorded on a LC TOF (ES) apparatus. The aryl triflates were synthesized accoridng to known procedures [7].

General Procedure for the Ni(0)-Catalyzed Heck Reaction

Aryl triflate (1 equiv), DPPF (0.05 equiv), Cy_2NMe (3 equiv), butyl vinyl ether (4 equiv) and $Ni(COD)_2$ (0.05 eq) were dissolved in dioxane (3 mL) and the sample vial was fitted with a Teflon sealed screwcap and removed from the glovebox. The reaction mixture was stirred at 100 °C for 18 h. Hydrolysis was performed adding 6 M HCl (3 mL) to the reaction mixture and stirred at room temperature for 1 h. Diethyl ether was added and the crude reaction mixture was washed twice with water and once with brine, dried over anhydrous $MgSO_4$ and filtrated. Upon concentration in vacuo the crude product was purified by colomn chromotography.

1-([1,1′-Biphenyl]-4-yl)ethanone (Table 6.1)

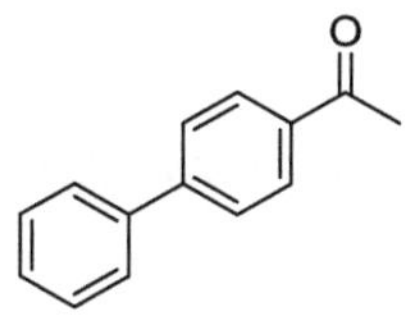

Flash chromatography using pentane:CH_2Cl_2 1:1 as eluent resulting in 26.4 mg (90 % yield) of the title product obtained as a colorless solid. ^{1}H NMR (400 MHz, $CDCl_3$) δ (ppm) 8.04 (d, 2H, J = 8.6 Hz), 7.69 (d, 2H, J = 8.6 Hz), 7.64-7.62 (m, 2H), 7.49-7.46 (m, 2H), 7.43-7.39 (m, 1H), 2.64 (s, 3H). ^{13}C NMR (100 MHz, $CDCl_3$) δ (ppm) 197.8, 145.9, 140.0, 136.0, 129.1, 129.0, 128.3, 127.4, 127.3, 26.8. HRMS $C_{14}H_{12}O$ [M+Na$^+$]; calculated 219.0786, found 219.0782.

1-([1,1′-Biphenyl]-2-yl)ethanone (Table 6.2, Entry 2)

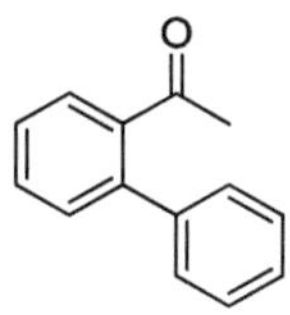

Flash chromatography using pentane:CH_2Cl_2 1:1 –> CH_2Cl_2 as eluent resulting in 86.4 mg (88 % yield) of the title product obtained as a colorless oil. ^{1}H NMR (400 MHz, $CDCl_3$) δ (ppm) 7.56 (dd, 1H, J = 1.2, 7.5 Hz), 7.51 (dt, 1H, J = 1.4, 7.5 Hz),

7.45–7.34 (m, 7H), 2.01 (s, 3H). ^{13}C NMR (100 MHz, $CDCl_3$) δ (ppm) 204.8, 140.9, 140.8, 140.5, 130.8, 130.3, 128.9, 128.7, 127.9, 127.9, 127.5, 30.5. HRMS $C_{14}H_{12}O$ [M+Na$^+$]; calculated 219.0786, found 219.0783.

1-(3-(Trifluoromethyl)phenyl)ethanone (Table 6.2, Entry 9)

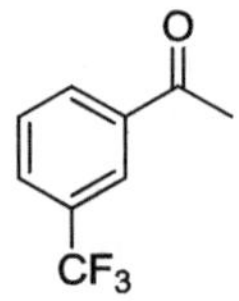

Flash chromatography using pentane:CH_2Cl_2 3:2 as eluent resulting in 75.8 mg (81 % yield) of the title product obtained as a colorless oil. ^{1}H NMR (400 MHz, $CDCl_3$) δ (ppm) 8.21 (s, 1H), 8.14 (d, 1H, $J = 7.8$ Hz), 7.49 (d, 1H, $J = 7.8$), 7.62 (t, 1H, $J = 7.8$ Hz), 2.65 (s, 3H). ^{13}C NMR (100 MHz, $CDCl_3$) δ (ppm) 196.7, 137.7, 131.6, 131.3, 129.7, 129.4, 125.2, 123.8 (q, J = 270.5 Hz), 26.7. ^{19}F NMR (376 MHz, $CDCl_3$) δ (ppm) -63.3. GCMS $C_9H_7F_3O$ [M]; calculated 118, found 118.

(E)-1-(4-((2-Methoxyphenyl)diazenyl)phenyl)ethanone (Table 6.2, Entry 12)

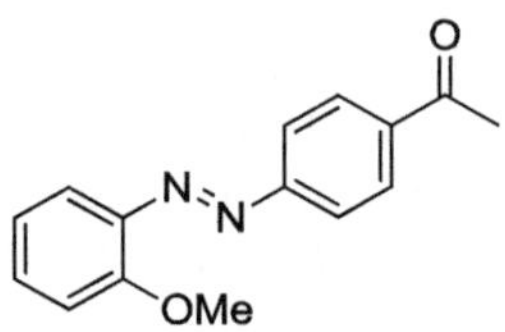

Flash chromatography using ethyl acetate:pentane 1:3 as eluent resulting in 95.4 mg (75 % yield) of the title product obtained as a red solid. ^{1}H NMR (400 MHz, $CDCl_3$) δ (ppm) 8.08 (d, 2H, J = 8.89 Hz), 7.94 (d, 2H, J = 8.08 Hz) 7.69 (dd, 1H, J = 1.80, 7.77 Hz), 7.47 (dt, 1H, J = 1.80, 8.89 Hz), 7.10 (dd, 1H, J = 0.52, 8.89 Hz), 7.02 (dt, 1H, J = 0.52, 7.98 Hz), 4.02 (s, 3H), 2.63 (s, 3H). ^{13}C NMR (100 MHz, $CDCl_3$) δ (ppm) 197.8, 157.7, 155.8, 142.4, 138.3, 133.7, 129.6, 123.2, 121.0, 117.1, 118.1, 56.5, 27.1; HRMS $C_{15}H_{14}N_2O_2$ [M+Na$^+$]; calculated 277.0953, found 277.0955.

1-(Benzo[d][1,3]dioxol-5-yl)ethanone (Table 6.2, Entry 8)

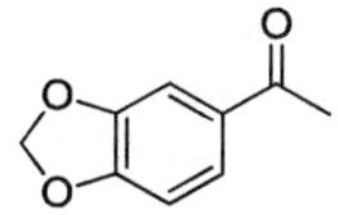

Flash chromatography using CH_2Cl_2 as eluent resulting in 62.9 mg (77 % yield) of the desired product obtained as a colorless solid. ^1H NMR (400 MHz, $CDCl_3$) δ (ppm) 7.53 (d, 1H, $J = 8.1$ Hz), 7.41 (s, 1H), 6.83 (d, 1H, $J = 8.1$ Hz), 8.02 (s, 2H), 2.52 (s, 3H). ^{13}C NMR (100 MHz, $CDCl_3$) δ (ppm) 196.2, 151.9, 148.3, 132.2, 124.8, 108.0, 107.9, 101.9, 26.5. HRMS $C_9H_8O_3$ [M+Na$^+$]; calculated 187.0371, found 187.0375.

7-Acetyl-2H-chromen-2-one (Table 6.2, Entry 16)

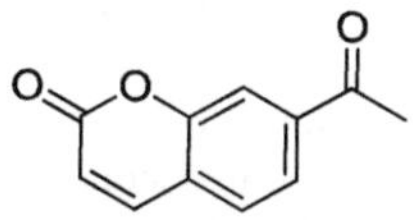

Flash chromatography using diethyl ether:pentane:CH_2Cl_2 2:2:1 as eluent resulting in 73.0 mg (78 % yield) of the title product obtained as a colorless solid. ^1H NMR (400 MHz, $CDCl_3$) δ (ppm) 7.82-7.79 (m, 2H), 7.73 (d, 1H, $J = 9.5$ Hz), 7.56 (d, 1H, $J = 7.9$ Hz), 6.48 (d, 1H, $J = 9.5$ Hz), 2.61 (s, 3H). ^{13}C NMR (100 MHz, $CDCl_3$) δ (ppm) 196.5, 160.1, 153.9, 142.5, 139.3, 128.3, 122.2, 118.9, 116.7, 26.9. HRMS $C_{11}H_{18}O_3$ [M+Na$^+$]; calculated 211.0371, found 211.0366.

Computational Details

All DFT calculations were carried out using the Jaguar program, [8] with the B3LYP hybrid functional [9] and the LACVP* basis set [10]. Complexes were optimized in vacuo and subjected to vibrational analysis to validate the nature of the stationary point as ground states or transition states, respectively. Solution phase energies were obtained by single-point calculations on the gas phase structure. The singlet-triplet gap for the investigated Ni complexes was examined by continually running single point energy calculations on the optimized singlet structures. In all cases, the triplet was found to be higher in energy.

The solvent was modeled by Jaguar's Poisson-Boltzmann method [11] with parameters describing THF ($\varepsilon = 7.6$, probe radius = 2.52 Å) which is a competent solvent for the title reaction, and computationally more feasible than the experimentally preferred dioxane. The solution phase Gibbs free energy was estimated by adding the gas phase vibrational contribution to the single-point solution phase energy. A dispersion correction term was calculated and added to give the total energy [12]. All transition states were refined by eigenmode following and verified by the presence of only one imaginary frequence. Visual inspection backed up that the eigenvector corresponded well to the formation of the correct bond or torsional rotation and QRC calculations were used to confirm this [13].

References

1. Hansen AL, Ebran J-P, Ahlquist M, Norrby P-O, Skrydstrup T (2006) Angew Chem Int Ed 45:3349
2. von Roman U, Ruhdorfer J, Knorr R (1993) Synthesis 985
3. Op den Brouw PM, Laarhoven WH (1982) J Royal Neth Chem Soc 101:58
4. Chowdhury S, Roy S (1997) J Org Chem 62:199
5. Ogata T,Hartwig JF (2008) J Am Chem Soc 130:13848
6. Gottlieb HE, Kotlyar V, Nudelman A (1997) J Org Chem 62:7512
7. Kwong FK, Lai CW, Chan KS (2001) J Am Chem Soc 123:8864
8. Jaguar v. 7.7 by Schrodinger LLC, New York, NY (2010) For documentation and current version of the program, see www.schrodinger.com
9. Becke AD (1988) Phys Rev A 38:3098–3100
10. LACVP basis sets use 6-31G for main group elements and the Hay-Wadt ECP for Ni: Hay PJ, Wadt WR (1985) J Chem Phys 82:299
11. Marten B, Kim K, Cortis C, Friesner RA, Murphy RB, Ringnalda MN, Sitkoff D, Honig B (1996) J Phys Chem Us 100:11775
12. Grimme S, Antony J, Ehrlich S, Krieg H (2010) J Chem Phys 132
13. Goodman JM, Silva MA (2003) Tetrahedron Letters 44:8233

MIX
Papier aus verantwortungsvollen Quellen
Paper from responsible sources
FSC® C105338

If you have any concerns about our products,
you can contact us on
ProductSafety@springernature.com

In case Publisher is established outside the EU,
the EU authorized representative is:
Springer Nature Customer Service Center GmbH
Europaplatz 3, 69115 Heidelberg, Germany

Printed by Libri Plureos GmbH
in Hamburg, Germany